THE FLOWER AND THE BEE

Torch-Lily or Flameflower. *Kniphofia aloides*

A bird-flower from South Africa widely cultivated in North America

THE
FLOWER AND THE BEE

PLANT LIFE AND POLLINATION

BY

JOHN H. LOVELL

BOTANICAL EDITOR OF THE "A B C OF BEE CULTURE"

ILLUSTRATED FROM PHOTOGRAPHS BY THE AUTHOR

NEW YORK
CHARLES SCRIBNER'S SONS
1918

PREFACE

MANY books have been published for the avowed purpose of making known our wild flowers. For botanical students there are the technical floras, and for those not familiar with botanical nomenclature there have been provided various artificial keys to our common trees, shrubs, and herbaceous plants. Formerly the study of botany consisted chiefly in "the mental gymnastics" of analyzing flowers to determine their names, but this is very far indeed from knowing them. The identification of a species should be regarded merely as an introduction and the beginning of a friendship long to be continued. The present volume treats of plants alive and in the midst of their home surroundings. There is no more attractive subject for investigation than the manifold ways in which flowers have solved their life problems, and have made use of both inanimate and animate agencies as pollen-carriers. The practical value of such observations cannot be easily overestimated, since they make the bee-keeper familiar with the resources of the honey flora, and save the fruit-grower from great disappointment and loss.

During many years the author has devoted a great amount of time to the observation of the life-relations of flowers. The forms and functions of the floral members have been carefully studied, and innumerable hours have been given to watching the behavior of the insect visitors and collecting them. For much of his time the field naturalist must be content with the companionship of Nature.

"The flower that on the lonely hillside grows
Expects me there when Spring its bloom has given;
And many a tree and bush my wandering knows."

The most important of insect pollinators is the honey-bee, and in order to become familiar with its economy I long since became a practical bee-keeper. Thus I have been able to approach the science of flower ecology from three different points of vantage. It will be evident in the following pages that the botanist, entomologist, or apiarist, who studies only one phase of this subject must necessarily obtain only a partial and imperfect view.

As a result of the supercritical spirit which has been abroad in the scientific world during the past generation most of the older biological theories have been called in question, and many new and bizarre suppositions have been advanced. The experience of the author has convinced him of the efficacy of natural selection in the evolution of flowers, of the advantages of cross-fertilization, and of the inheritance of acquired characters. In the absence of insect visits there is no satisfactory evidence that conspicuous flowers and their adaptations would ever have been developed. So highly improbable and quixotic are some of these new theories that biologists should not forget that common sense is an important factor in the interpretation of nature. Let us not mistake for giants what in reality are only windmills.

While technical terms have been avoided, great care has been taken to secure accuracy of statement. Such difficulties as occur, it is believed, will prove stimulating rather than discouraging. Most of the photographs are natural size, except

PREFACE

in the case of a few of the larger flower-clusters, and have been taken on panchromatic plates to preserve in monochrome the proper color values. A small stop and a long exposure have been employed to secure details. By the aid of the figures the reader will be able to follow the descriptions with nearly as much ease as with the natural flowers. It is hoped that the volume may inspire in those who have been content in merely learning the names of flowers, a desire to know more of their biology, both past and present, and that in the schools, on the farms, and among all students of nature it may awaken a greater interest in the study of plant life.

Much of the material used in the preparation of the following chapters has appeared, partly in more technical form, and partly in popular articles in the *American Naturalist*, the *Journal of Animal Behavior*, the *Scientific Monthly*, *Entomological News*, *Psyche*, *Gleanings in Bee Culture*, the *A B C of Bee Culture*, *A Cyclopedia of Everything Pertaining to the Honey-Bee*, the *American Bee Journal*, the *Lewiston Journal*, etc. Without this long preliminary preparation they would probably never have been written. Grateful acknowledgment is also made to my wife for assistance in statistical and entomological work.

JOHN H. LOVELL.

CONTENTS

xi

ILLUSTRATIONS

ILLUSTRATIONS

xiv

ILLUSTRATIONS

ILLUSTRATIONS

ILLUSTRATIONS

xvii

THE
FLOWER AND THE BEE

CHAPTER I

FLOWERS AND HUMANITY

D URING the past half-century we have been learning
as never before in the history of the human race the
great importance of keeping in close contact with nature. Our future health and prosperity depend upon our love
for the soil and its productions. The Greek fable which tells
how the giant Antæus, while wrestling with Hercules, never
failed to renew his strength whenever he touched his mother
earth, will always be true of man, both physically and morally.
Of all natural productions, there is none so well adapted for
maintaining an intimate communion with nature as the cultivation and study of flowers. Whoever plants a flower-
garden, benefits not only himself but his whole village. If
the human brain is the most wonderful production of evolution, as Haeckel asserts, flowers are the most beautiful; and,
says William Winter, ' the ministry of beauty is the important
influence upon society that can never fail.'

There is a fascination about an old garden, indeed, that few
can resist. I am glad that the first botanical garden in America,
which was planted by John Bartram, the first American bota-

nist, is still preserved as a public park by the city of Philadelphia. It contained a great variety of shrubs and trees, as well as herbaceous plants, raised from seeds and roots collected during his numerous journeys and received from his European correspondents. There was a greenhouse built by Bartram himself, over the door of which were inscribed the lines:

"Slave to no sect, who takes no private road,
    But looks through nature up to nature's God."

One of Bartram's correspondents was Peter Collinson, a London merchant, who had a choice garden, the pride of his life, at Mill Hill, where he skilfully cultivated rare species of plants received from the colonies. In one of his letters to Bartram he exclaims: "Oh, Botany! delightfullest of all the sciences! there is no end to thy gratifications." No one who has not experienced it can realize how intense is the enjoyment of watching the blooming of plants. A short time before his death Keats told his friend Severn that he thought that his intensest pleasure in life had been to watch the growth of flowers.

Among children the love of flowers is universal. Says one writer: "I think I never knew a child that did not love flowers. Many children are passionately fond of them, but I never knew a child indifferent to them." Children and flowers! Flowers and children! Surely they are the two chief sources of human happiness! Says Donald G. Mitchell: "Flowers and children are of near kin. I love to associate them, and to win the children to a love of the flowers." I know of a little lad to whom the succession of flowers brings one of the chief joys of the year. With what delight he watches for each blossom in spring, and how eagerly he tells of the treasure he has found! Here is a pleasure that is free to all, and yet is greater than any money

2

can buy. When it is remembered that in many cities there are children who have never seen a buttercup, the value of maintaining flower-gardens in city squares and in every available spot cannot be overestimated. Let us hope that the time will speedily come when every child, both at home and by means of the school-garden, will be taught the fundamental facts of plant life, not alone for the practical advantages to be gained, great as these are, but that they may have through life a never-failing resource, in the pursuit of which they can always find happiness and contentment.

But great as is the pleasure that flowers bestow, it is far from being the only benefit received from them. It has been rightly said that "nothing teaches us so much in this world as flowers, if we will only watch them, understand the messages they exhale, and profit by them. Every lesson in life is taught by the flowers; every message to the human heart is carried in them." Nor is the time devoted by the professional or laboring man to the investigation of flowers wasted, even from a practical point of view. Charles Kingsley has forcibly described the helpfulness of such studies:

"I know of few studies to compare with natural history; with the search for the most beautiful and curious productions of nature amid her loveliest scenery, and in her freshest atmosphere. I have known again and again working men who, in the midst of smoky cities, have kept their bodies, their minds, and their hearts healthy and pure by going out into the country at odd hours and making collections of fossils, plants, insects, birds, or some other objects of natural history; and I doubt not that such will be the case with some of my readers."

"Supposing that any of you, learning a little sound natural history, should abide here in Britain to your life's end, and observe nothing but the hedgerow plants: he would find that there is much more to be seen in those mere hedgerow plants than he fancies now. . . . Suppose that he learned something of this, but nothing of aught else. Would he have gained no solid wisdom? He would be a stupider man than I have a right to believe any of my readers to be, if he had not gained thereby somewhat of the most valuable of treasures, namely, that inductive habit of mind—that power of judging fairly of facts, without which no good or lasting work will be done, whether in physical science, in politics, in philosophy, in philology, or in history."

"Take my advice for yourselves, dear readers, and for your children after you; for, believe me, I am showing you the way to true and useful, and, therefore, to just and deserved power. I am showing you the way to become members of what I trust will be—what I am sure ought to be—the aristocracy of the future."

Many farmers and fruit-growers too readily assume that a knowledge of the life histories of flowers can never aid them in getting a better livelihood. There could be no greater mistake. The larger part of our cultivated fruits are either partially or wholly self-sterile, and in the absence of bees and other pollinating insects remain either entirely barren or largely unproductive. Fruit-culture on the extensive scale in practice at the present time would be impossible without the domestic bee; and it is estimated by Phillips that bee-keeping annually adds indirectly more to the resources of the country by flower-pollination than by the sale of honey and wax. In sections where immense orchards cover many square miles of territory, the wild insects are wholly inadequate to pollinate the great expanse of bloom, and numerous apiaries must be maintained to obtain the best results. An intimate knowledge of the way fruits and vegetables are pollinated is, therefore, of inestimable value to the agriculturist.

To the bee-keeper also familiarity with the honey flora is indispensable, and may determine the failure or success of his efforts. Nectar-bearing plants may be abundant in one locality, and comparatively rare a few miles away. Too often if there is a small surplus of honey, he does not know whether the fault is with the bees or with the honey flora. Unfortunately, to many an apiarist the wild flowers always remain strangers.

> "Primroses by the river's brim
> Dicotyledons are to him,
> And they are nothing more."

Then, again, there are some bee-keepers who appear to look upon flowers as created or evolved solely for the benefit of bee-culture. They are slow to realize that there are blossoms which are nectarless, or which contain nectar which is inaccessible to honey-bees. Accordingly we find from time to time, bird-flowers, bumblebee-flowers, butterfly-flowers, and moth-flowers, pollen-flowers, and wind-pollinated flowers reported as excellent honey plants. That a flower should produce nectar plentifully, but at the bottom of a tube so long that honey-bees cannot reach it, seems to them an evidence, as a Yankee once remarked, that "Providence was kind, but careless." Nature fashioned the wild flowers before the human race appeared upon the earth, and they would not have been one whit different to-day had the appearance of mankind been deferred to some distant future.

Undoubtedly the influence of flowers upon the development of the human race has been both profound and far-reaching. So intimately do they enter into every phase of life, and so eloquently do they express every emotion, that it was long believed that their bright colors, sweet odors, and varied forms were created solely for the benefit of man. We cannot imagine what this world would have been without them, or estimate the enjoyment that would have been lost, or the power for good that would have been forever missing; but we know that humanity would have been less perfect than it is to-day. And the loss of conspicuous flowers is not inconceivable, for their development is correlated with insect visits, and in their absence our flora would have been composed chiefly of small, green or dull-colored blossoms, similar to those of the grasses and sedges and of thousands of other plants, which are wind-pollinated, and are usually passed by almost unnoticed.

That flowers act strongly upon the imagination is shown by the myths of the Greeks, and the poetry of all nations. Even the ruder songs of the primitive northern nations, according to Humboldt, were influenced by the forms of plants. Of the relations of flowers to humanity, the poet is the true interpreter, not the man of science. He alone, as Longfellow has said, is qualified to unfold the bright and glorious revelations and the wondrous and manifold truths written in these stars of earth.

"And the poet, faithful and far-seeing,
  Sees, alike in stars and flowers, a part
Of the selfsame universal being
  Which is throbbing in his brain and heart.

Brilliant hopes, all woven in gorgeous tissues,
  Flaunting gaily in the light;
Large desires, with most uncertain issues,
  Tender wishes blossoming at night.

These in flowers and men are more than seeming;
  Workings are they of the selfsame powers
Which the poet in no idle dreaming
  Seeth in himself and in the flowers.

In all places, then, and in all seasons,
  Flowers expand their light and soul-like wings,
Teaching us by most persuasive reasons
  How akin they are to human things."

While an examination of the poetry which has been written on flowers in all ages would teach many valuable lessons, we

must be content to quote three verses from Leigh Hunt's
"Songs of the Flowers," in which he surpasses all other poets
in his description of the life of flowers and their relation to humanity. From the point of view of the naturalist this is the
most remarkable poem on flowers in any language, "fathoming,"
says Hamilton W. Mabie, "the very soul of flowers." "No
poet in this nor in many a generation past has said a sweeter
or more haunting word for the flowers."

> "We are the sweet flowers,
>   Born of sunny showers,
> Think, whene'er you see us, what beauty saith:
>   Utterance mute and bright
>   Of some unknown delight,
> We fill the air with pleasure, by our simple breath:
>   All who see us, love us;
>   We befit all places;
> Unto sorrow we give smiles; and unto graces, graces.

>   See, and scorn all duller
>   Taste, how Heav'n color lover,
> How great Nature, clearly joys in red and green;
>   What sweet thoughts she thinks
>   Of violets and pinks,
> And a thousand flashing hues made solely to be seen;
>   See her whitest lilies
>   Chill the silver showers,
> And what red mouth has her rose, the woman of the flowers.

>   Think of all these treasures,
>   Matchless works and pleasures,
> Everyone a marvel, more than thought can say;
>   Then think in what bright show'rs
>   We thicken fields and bowers,
> And with what heaps of sweetness half wanton May.
>   Think of the mossy forest
>   By the bee-birds haunted,
> And all those Amazonian plains, lone lying as enchanted."

CHAPTER II

THE DISCOVERY OF THE SECRET OF FLOWERS

THE human race has long assumed (being the only organism at liberty to place upon itself its own valuation) that it occupies a position of fictitious importance in the universe. It was a current maxim in the Middle Ages that man was the measure of all things. The world and its inhabitants, so ran this pleasant myth, was created a few thousand years ago, solely to provide him with a congenial place of abode; and, because of his paramount importance, was placed in the centre of the heavens. Not a little ingenious (and to-day amusing) speculation was expended in an effort to explain how natural cataclasms and noxious animals and plants were disguised blessings; but that such was the fact, no doubt was permitted to exist. From these modest pretensions we have been receding for some centuries with much hesitation and reluctance. Perhaps the close of another hundred years will see them abandoned altogether, and humanity willing to admit that it is a part of nature, not outside and above her.

So long as these teachings prevailed it was very naturally a popular notion that the bright colors of flowers were of no importance except as they gave human pleasure. Much superfluous pity was wasted on those blossoms which, to use the words of the poet Gray, blushed unseen and wasted their sweetness on the desert air. Only a few years ago a similar sentiment was expressed by the editor of one of our popular magazines: "There was apparently no particular reason why the earth, at the time of Adam, should have been literally strewn with blossoms. They were of no particular use; there was only one man to see them."

This same idea is again repeated in Emerson's beautiful lines:

THE RHODORA

"In May, when sea-winds pierced our solitudes
I found the fresh rhodora in the woods,
Spreading its leafless blooms in a damp nook;
To please the desert and the sluggish brook:
The purple petals fallen in the pool,
Made the black waters with their beauty gay:
Here might the red-bird come his plumes to cool,
And court the flower that cheapens his array.
Rhodora! if the sages ask thee why
This charm is wasted on the marsh and sky,
Dear, tell them that, if eyes were made for seeing,
Then beauty is its own excuse for being."

It would seem never to have occurred to poet, editor, or philosopher that the beautiful hues of flowers might be useful to the plants producing them.

It was a German pastor, Christian Conrad Sprengel, at the close of the eighteenth century, who first pointed out the true significance of conspicuous flowers. His book, now a botanical classic, attracted but little attention; his publisher did not even send him a copy of it, and in discouragement he did not publish the second volume, but turned from the study of plants to that of languages. The title of the work, *The Secret of Nature in the Form and Fertilization of Flowers Discovered*, affords us the pleasure of knowing that he rightly estimated the importance of his observations. Sprengel clearly states, as is now well established, that the bright hues of flowers serve as signals to attract the attention of nectar-loving insects flying near by. He was led to this conclusion very fitly by the study of *Myosotis,* or the forget-me-not. He has not been forgotten. His name and theory were rescued from obscurity by Darwin; his book a few years ago was republished at Leipsic, and is now universally recognized, says Mueller, as having "struck out a new path in botanical science."

Sprengel was convinced that the wise Framer of nature had not produced a single hair without a definite purpose, and he examined a great many flowers for the purpose of learning the meaning of their forms and the arrangement of their parts. The salver-formed flower of the forget-me-not is sky-blue with a yellow eye. "While studying the flower of *Myosotis* I was struck," he says, "by the yellow ring which surrounds the opening of the corolla tube, and which is beautifully conspicuous against the sky-blue of the limb." (Fig. 1.) "Might not, I thought, this circumstance also have some reference to insects? Might not nature have especially colored this ring, to the end that it might show insects the way to the nectar-reservoir?" On further observation he found that the entrances to many other flowers were marked with spots, lines, and dots differently colored from the rest of the corolla. These marks he called "nectar guides." "If the particular color of one part of a flower," he rightly inferred, "serves to enable an insect, which has settled on the flower, easily to find the right way to the nectar, then the general color of the corolla is serviceable in rendering the flowers provided with it conspicuous even from afar to the eyes of insects that hover around in the air in search of food."

Sprengel decided that flowers secrete nectar for the sake of attracting insects, and that it is protected by hairs or nectaries in order that they may enjoy it pure and unspoiled. At first he thought that the flowers received no service in return; but he soon observed that the guests pollinated the flowers. He

Fig. 1. Forget-Me-Not. *Myosotis scorpioides*
Blue flower with yellow eye

even noticed the frequent occurrence of cross-pollination, and remarks that "it seems that nature is unwilling that any flower should be fertilized by its own pollen." He described the manner in which some five hundred flowers are pollinated; but as he knew little about insects he did not pay much attention to the different kinds of visitors.

But while Sprengel had learned the secret of flowers and knew that their colors, odors, and forms were not useless characters, he failed to discover why cross-pollination is beneficial; and this omission, as Mueller has remarked, was for several generations fatal to his work. In 1841 Robert Brown, an eccentric English botanist of great learning, advised Darwin to read Sprengel's book. "It may be doubted," says Francis Darwin, "whether Robert Brown ever planted a more beautiful seed than putting such a book into such hands." Thus is the torch of learning, shining with ever-increasing effulgence, handed on from one investigator to another. Darwin was already engaged in studying British orchids, of which he wrote to Bentham: "They are wonderful creatures, these orchids." His interest in the structure and pollination of these curious plants was greatly increased by reading what the old German pastor had done. Darwin soon discovered that frequent crosses increase both the vigor and productiveness of the stock, and that an occasional cross is indispensable. The principal agents which nature employs for this purpose are insects, birds, wind, and water. So impressed was Darwin with the importance of cross-fertilization that he closed his famous book on orchids, which marks the next great epoch in flower ecology, with the words: "Nature abhors perpetual self-fertilization." "The charm," says Mueller, "was now broken, and the value of Sprengel's work was at once recognized." "The merits of poor old Sprengel," says Darwin in his autobiography, "so long overlooked, are now fully recognized many years after his death."

In 1866 Darwin's *Origin of Species* and his book on orchids were read by Hermann Mueller, a young teacher at Lippstadt, Germany, who thenceforth enthusiastically devoted the rest of his life to the study of the pollination of flowers. Many other investigators were also stimulated by these epoch-making books to study the charming problems of floral structure, as Delpino in Italy, Axel in Sweden, Hildebrand in Germany, Asa Gray in North America, and Fritz Mueller in South America. But they were all easily surpassed by Hermann Mueller, who is still regarded as the foremost of floroecologists. In Thuringia and in the Alps he examined many hundreds of blossoms and recorded the visits of insects by thousands. He was the first to collect and publish lists of flower-pollinators on an extensive scale, and the biology of flowers may thus be said in its broadest sense to have been established by him. Never since has this branch of botany been cultivated with equal success. His book *The Fertilization of Flowers* ranks with the works of Sprengel and Darwin, and marks the third great epoch in the history of flower ecology. (Fig. 2.)

Hermann and Fritz Mueller were the sons of the Evangelical pastor at Mühlberg, in Thuringia. Fritz was born in 1822 and Hermann in 1829. Hermann was deeply attached to his native land, and often in his later life referred to it as "his dear Thuringia." There with Fritz he explored the fields and streams, and under the influence of his studious elder brother his love for the plant world was awakened.

After preparing himself for a teacher and also studying medicine Fritz emigrated to Brazil, where he settled at Blumenau as a farmer. Afterward he went to the Lyceum at Desterro, but, on being driven from office by the Jesuits, he returned to Blumenau and became travelling naturalist for the province of Santa Catharina. From his new home he rendered invaluable service to his fatherland by frequently communicating the results of his scientific researches and important observations. After his brother Hermann's death he wrote: "With Hermann I have during the last twenty years exchanged one letter regularly each month, nor did either of us wait for the answer to our last letter before writing again." By this lively exchange of ideas, which related chiefly to their investigations in natural history, each incited the other to greater efforts. On the expulsion of the Emperor Dom Pedro, Fritz was deprived of his office and pay without explanation, and letters were often not delivered to his address. During a battle near Blumenau the revolutionists robbed him of a part of his property, imprisoned him for eight days, and he escaped with his life only by a fortunate accident. He died in 1897; his best-known work was *Für Darwin*, or *Facts for Darwin*. Two of his grandsons inherit his love for nature.

The younger brother, Hermann, graduated from the gymnasium at Erfurt in 1847. During his spare time he found youthful employment in studying the floral wealth in the environs of that city. At the University of Halle, and later at Berlin, geology became his favorite pursuit. Two journeys to the Alps awakened in him an appreciation of the rich flora and fauna of these mountains. In 1855 he became teacher in the newly established Realschule in Lippstadt, and ten years later he was appointed uppermaster, a position he retained until the close of his life. His first book, *The Fertilization of Flowers*, won the praise of Charles Darwin and has had a world-wide usefulness as a work of reference on flower-pollination. It is illustrated by many excellent woodcuts, the drawings for which were made by Mueller himself. It contains descriptions of the floral mechanisms of many plant species, with lists of their insect visitors. It will give some idea of the immense amount of labor involved in its preparation to state that

FIG. 2. Hermann Mueller

5,231 visits to flowers by 843 different kinds of anthophilous, or flower-visiting, insects are recorded.

Mueller had never forgotten his earlier delightful journeys among the Alps nor its rich and brilliantly colored flora. For six summers he continued with great diligence to investigate its flowers, and the result was his second great work, entitled

THE FLOWER AND THE BEE

Alpenblumen, or the *Flowers of the Alps*. Here are enumerated 5,711 visits by 841 species of anthophilous insects. It is impossible to read this account of the mysteries of the floral world in high altitudes without longing to visit the scene of his investigations. The short summers, the rapid (not to say impetuous) advance of vegetation, the simultaneous blooming of many species, the brilliant hues, the wealth of insects, and especially the great abundance of butterflies, against a background of snowy summits, form a most enticing picture. Mueller published a third book on flowers, besides many shorter papers.

Hermann died very suddenly, in 1883, while studying the flowers of the Tyrol. He was travelling in part for the benefit of his health, but he was without any premonition of his fate. On the day of his death he had written a long letter to his son at Lippstadt, and his valise was packed for his departure the next morning. Suddenly, on the evening of the 25th of August, a pulmonary attack closed his useful life. It was fitting that a life devoted to the study of highland flowers should come to its close among them. "He is not dead," says his biographer, Ludwig; "he lives, and will live so long as a flower enraptures the eye of an investigator. His bright spirit will live and, we hope, like that of his teacher and friend Darwin, long be a light on the way to truth in the heart of nature."

Since Mueller's death the most important undertaking has been Knuth's *Handbook of Flower Pollination*, an encyclopædic work in three volumes, giving a complete summary with a bibliography of some four thousand titles of everything that has been done in floroecology up to the beginning of the present century. It was planned and the first two volumes brought out by Paul Knuth, and after his untimely death, at forty years of age, the third volume was completed by Ernst

16

Loew, of Berlin. An excellent translation by J. Ainsworth Davis of the first and second volumes has appeared.

To-day there are very few investigators engaged in studying the life histories of flowers. In North America they number less than half a dozen. Most observers are content to restrict their attention to the botanical side of the subject, and ignore the great company of pollinators. Even Charles Darwin and Anton Kerner, whose writings still remain an ever-inspiring source of information, gave little heed to the ways of the insect guests. The reason for this is not far to seek. To collect and prepare lists of the visitors, and to observe their behavior, requires so enormous an amount of time, labor, and patience that the opportunity is possible to very few people. Suppose that a flower is in bloom for two weeks, then, on every calm, bright day many hours must be devoted to the work, for the guests at the beginning of blooming-time may differ from those at its close. There follows the almost insuperable task, at least in America, of determining the names of the captured insects. With the exception of the butterflies we have no manuals of the different orders, and the literature is in a truly chaotic condition, many papers not being obtainable at any price. It is noteworthy that each of the three or four more prominent investigators of floroecology in America has been compelled to work up the classification of the bees in his locality—a rather formidable undertaking in itself. So closely allied are the species and genera that no one can distinguish between them without a special knowledge of the group, which in its relations to flowers exceeds all others in importance.

But the value of an acquaintance with the insect visitors cannot easily be overestimated; for some species fly only in the spring, others only in the fall; some species visit only one kind of flower, others many kinds; some are most welcome

17

guests, others are mere robbers. I should never have dreamed
that the pretty, nodding pink blossoms of the twinflower
(*Linnæa borealis*), with its sweet vanilla-like fragrance, are in

FIG. 3. Bladderwort. *Utricularia vulgaris*
The two-lipped yellow flowers are pollinated by syrphid flies, both in Europe and America

our northern woodlands attractive to gnats alone. One afternoon a large bed of these delicate flowers was carefully observed,
and eight visitors were collected. On examination they were

found to belong to a single species of fly (*Empis rufescens*). Further observations show that in this locality this fly is probably the only guest. A burly bumblebee flew over the flowers without paying any attention to them.

Among aquatic plants living in fresh-water rivers is the bladderwort (*Utricularia vulgaris*). The whole plant is submerged; but at blooming-time a flower-stalk is thrust out of the water, which produces deeply two-lipped, bright yellow flowers. (Fig. 3.) I certainly expected to find it a favorite of bees. But after repeated observations I have collected on the flowers in Maine only the long-tongued syrphid fly (*Helophilus conostomus*). There is no way in which we can so easily learn the defects of flowers as to watch the behavior of insects upon them. No human eye can discover them so quickly. In a word, if we would fully understand the bright-hued floral edifices which so freely adorn the outdoor world we must study the *modus operandi* of their architects and builders.

CHAPTER III

FLOWERS POLLINATED BY THE WIND

"Let us have eyes to see
The new-old miracle;
If it befell
We viewed for the first time such wizardry,
Each budding leaf were past belief,
Ineffable."

EARLY spring in temperate North America has its own
peculiar charm. It is a period of conflict between two
seasons, ushered in by shrill winds and rushing waters.
Snow and ice still linger in the woods and cold days and frosty
nights are common; but warmth and light are continually
gaining. Among the many events characteristic of that old but
ever new miracle, the resurrection of plant life, the most noteworthy is the bursting into bloom suddenly in early spring of
whole forests of deciduous-leaved trees and shrubs before their
leaves have appeared. Many of them are familiar species as
the alders, birches, poplars, hazels, willows, hornbeams, walnuts,
hickories, beeches, elms, oaks, and chestnuts. A part produce
edible nuts, and a part are planted as avenue or ornamental
trees. Most people never know that they bloom at all. This
is partly because the flowers are small and dull-colored, and
partly because they come at a season of the year when they
are not expected. Why do the flowers expand before the leaves?
A few years ago no one could have answered this question.

When Louis Agassiz, the famous naturalist and the founder
of the Agassiz Museum at Cambridge, and Alexander Braun,
who afterward became a distinguished botanist, were school-

boys together at Carlsruhe, in 1827, Braun in one of his letters sent his friend some "nuts to pick," among which was the question: "Why do some plants blossom before they put forth leaves?" Years before Gilbert White, the naturalist of Selborne, had pondered over the same problem. "Why," he writes, "do some plants bloom in the very first dawnings of spring, some at midsummer, and some not till autumn? This circumstance is one of the wonders of creation, little noticed because a common occurrence—but it would be as difficult to be explained as the most stupendous phenomena in nature."

Difficult as the problem once seemed, there is no longer any mystery why the flowers of many forest-trees appear before their leaves. They are or were in time past pollinated by the wind, although the willows and maples have in comparatively recent years changed over to insect-pollination. At its best, wind-pollination is a very wasteful method of obtaining the advantages of cross-fertilization, and much of the pollen falls where it is of no benefit. It would clearly be almost a total failure in the case of the shrubs and trees enumerated, if their branches were covered with a dense foliage which intercepted the pollen. So the flower-buds are formed the preceding season, and begin to bloom in spring just as soon as the weather is warm enough. If Gilbert White were living to-day this phenomenon would excite his astonishment less and his admiration more.

The common alder is one of the commonest of New England shrubs, growing everywhere in swamps and wet land, and it is also an excellent example of wind-pollination. It blooms early in April, and, where the Mayflower is not found, is the true harbinger of a new season. (Fig. 4.)

"By the flowing river the alder catkins swing,
And the sweet song sparrow cries, 'Spring! it is spring.'"

21

The flowers of the alder are in catkins or aments. A catkin is a cylindrical flower-cluster composed of many small,

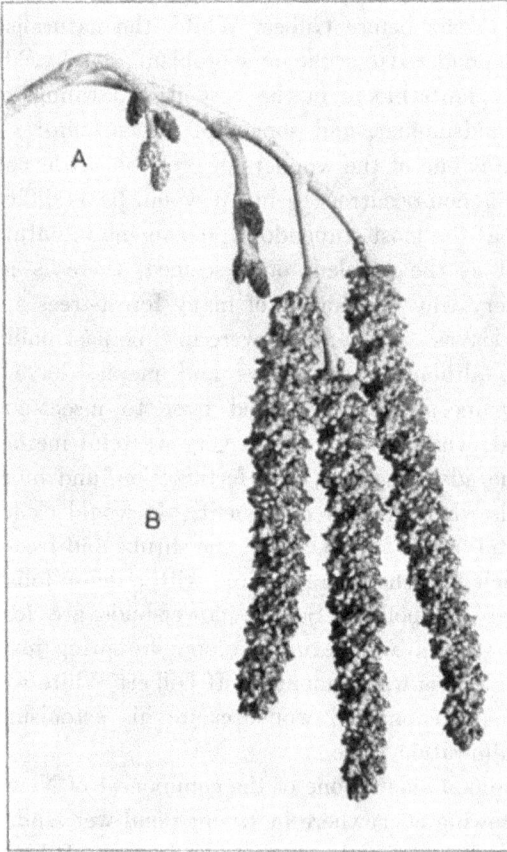

Fig. 4. Common Alder. *Alnus incana*
A, fertile catkins ; B, staminate catkins ; a wind-pollinated shrub

nearly sessile flowers, and numerous protecting scales. The stamens and pistils are always in different catkins. In a

staminate catkin of the alder I found by actual count 77 flowers
and 310 stamens; while a pistillate catkin contained 80 pistils.
A tassel of three or four staminate catkins occurs at the end of
a branch, while an inch or two higher up, where the pollen
cannot fall upon them, are one or two clusters of pistillate or
fruiting catkins. This arrangement facilitates cross-pollination. There are no allurements to attract insects, such as nectar, bright color, odor, or resting-places, for the wind is the
agent which carries the pollen. On a warm, clear day the
last of March or early in April, often before the swamps are
wholly free from ice, the anthers open and weigh down the
passing breeze with little clouds of pollen. Most of the pollen
would be wasted if the bushes were in full foliage. The honey-
bees obtain their earliest supply of pollen from the alders, and
in northern New England they are fortunate if their work is
not interrupted by a snow-storm. The seeds mature during
the summer, and the following winter are dropped upon the
drifting snow.

Later in April the willows begin to bloom, the familiar pussy-
willow (*Salix discolor*) leading the way. While no one ever
gathers the dark reddish-brown catkins of the alder, the pussy-
willows are very attractive and are often sold in New England
cities by street flower-venders, and used for decoration in
English churches on Palm Sunday. The staminate and pistillate catkins are borne on different trees or bushes so that
self-pollination is impossible. In a staminate catkin I counted
270 flowers, and in a pistillate catkin 142 flowers. The blossoms are sweet-scented, secrete nectar freely, and on a warm
day are sought by a great company of honey-bees, bumblebees, solitary bees, flies, and a few butterflies and beetles, from
which it is evident that they are to-day pollinated by insects.
But since the flowers are in catkins and "bare are the branches,

and cold is the air" when they bloom, we are safe in concluding that earlier in their history they were wind-pollinated; indeed, in Greenland anemophilous willows are still found. Salix is a very old genus, and fossil willow leaves occur in the rocks earlier than those of most modern plants. The transition in the manner of pollination has probably been brought about by bees, which visit the catkins in great numbers for pollen.

The willows are a bonanza to the bee-keeper, for they bloom at a time when the winter stores in the hives are beginning to fail, and are a great help in tiding over our cold, inclement springs. They furnish the first honey of the season, and with them begins the procession of the honey-plants, or plants valuable in the production of honey. A small surplus is sometimes obtained, which has a pleasant aromatic taste not unlike that of fruit-bloom.

Common forest trees and shrubs, which are catkin or ament-bearers (*Amentiferæ*) and are wind-pollinated, are the poplars, birches, hornbeams, bayberries, oaks, hazels, sweet-fern, walnuts, butternuts, and hickories. The staminate and pistillate catkins of the poplars are borne on different trees. When the weather is clear and dry the elastic anthers expel the pollen grains several inches into the air, and thus give them a fair start on their journey. The bayberry is also diœcious. (Fig. 5.) The white, yellow, and gray birches and the hornbeams are monœcious, or produce both kinds of catkins on the same trees. (Fig. 6.) The staminate flowers of the oaks, hazels, walnuts, and hickories are in long, drooping catkins, but the pistillate or fertile flowers are solitary or few in a cluster. The beech, elm, and ash are likewise wind-pollinated, but none of the flowers are in catkins. In the beech the pistillate flowers are in pairs, and the staminate blossoms in globose heads which droop downward on long, flexible stems. The flowers of the

FIG. 5. Bayberry. *Myrica carolinensis*

A, fertile catkins; *B*, sterile catkins, producing only pollen. A wind-pollinated shrub

elm are largely hermaphrodite or perfect, and self-pollination is to a great extent prevented by the stigmas maturing earlier than the anthers. Both the staminate and pistillate flowers

FIG. 6. Gray Birch. *Betula populifolia*
A, fertile catkin; B, staminate catkin. A wind-pollinated tree

of the ash are in clusters, and in our species are usually on different trees.

All of these genera possess certain contrivances or adaptations which tend to prevent self-pollination and to strongly favor cross-pollination. The stamens and pistils are often in

26

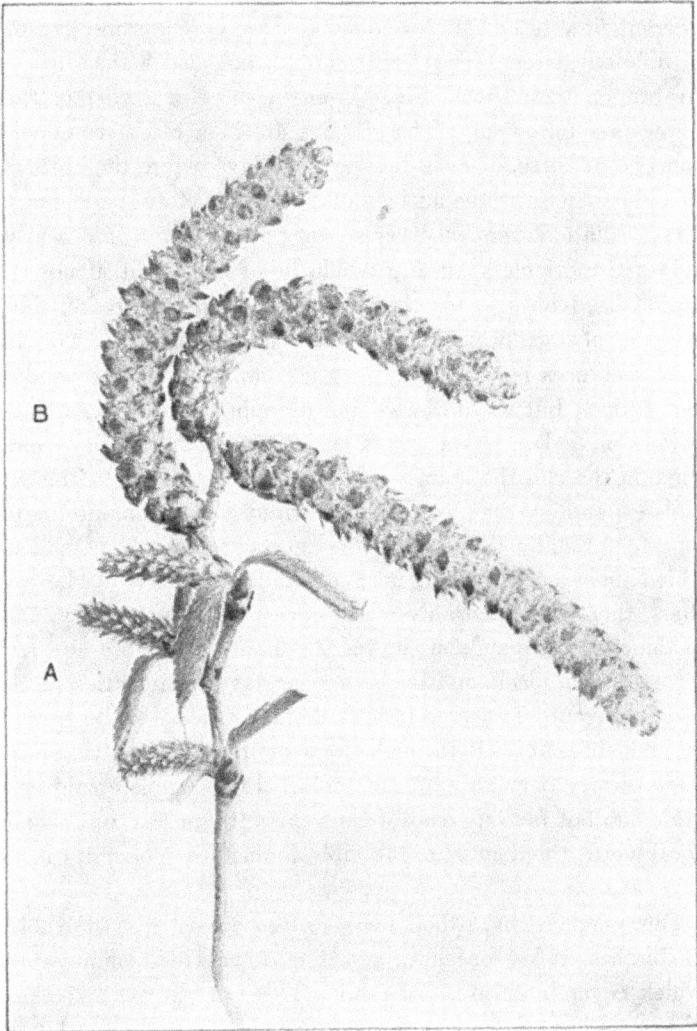

FIG. 7.   Yellow Birch.   *Betula lutea*
A, fertile catkins; *B*, staminate catkins.   A wind-pollinated tree

different flowers and flower-clusters either on the same tree or on different trees. The fertile flowers are also higher up on the branch than the sterile. Even in the elm a part of the flowers are unisexual. The stigmas are in a receptive condition two or three, or even four or five, days before the anthers open, or sometimes the anthers may mature before the stigmas. The pistillate flowers, whether solitary or in clusters, are nearly rigid and motionless, since it would be of no benefit, if not an actual disadvantage, for them to oscillate in the breeze; but there is always provision for shaking the anthers. This in most instances is very effectively accomplished by the pendulous catkin; but in the beech the drooping head of staminate flowers on a long stem sways easily to and fro in the wind, while in the elm the stamens double in length after the flower-buds expand, so that a slight air current causes the anthers to vibrate. Finally they all produce a copious supply of pollen, which must be protected from too much moisture. The anthers, therefore, open only when the air is warm and dry, and in the event of a sudden shower the fissure may close in a few seconds. An ideal condition is a clear day and a gentle breeze, which prevents the pollen from falling too quickly to the ground and widely diffuses it through the atmosphere, permitting it to settle slowly over an extensive area. In a complete calm the grains do not fall out readily from the anther-cells; but a high wind whirls them away too forcibly in one direction and causes excessive waste.

Our northern hardwood forests often consist almost wholly of birches, oaks, beeches, and the other trees enumerated, which cover large areas of land. There are literally acres of bloom and a numberless host of flowers. For the most part they are a delicate pale green, not easily distinguished from the newly expanding leaves; but the catkins of the yellow birch

Fig. 8. American Elm. *Ulmus americana*

A wind-pollinated tree with flowers in clusters

are a golden yellow, and the elms display great masses of reddish-purple blossoms. (Figs. 7 and 8.)

Two other immense groups of plants, which are anemophilous or wind-pollinated, are the grasses (*Gramineœ*) and the sedges (*Cyperaceœ*). There are some 3,000 species of sedges and 3,500 species of grasses; but great as is the number of species, their importance consists in the myriads of individuals which cover a large part of the earth's surface and provide most of the food material of the human race and herbivorous animals. To the grasses belong the edible cereals, corn, wheat, rye, barley, oats, rice, and millet. "Next to the importance," says Ingalls, "of the divine profusion of water, light, and air, those great physical facts which render existence possible, may be recorded the universal beneficence of grass. It is the type of our life, the emblem of our mortality. It bears no blazonry of bloom to charm the senses with fragrance, or splendor, but its homely hue is more enchanting than the lily or the rose. Should its harvest fail for a single year, famine would depopulate the earth."

Most grasses have perfect or hermaphrodite flowers, and self-pollination is largely prevented by the anthers and stigmas maturing at different times; but Indian corn is a familiar example of a grass with unisexual flowers. The spindles or staminate flower-clusters terminate the stalks and are borne well above the foliage, while the pistillate clusters (the ears in the silk) stand much lower down, where they are more likely to be cross-pollinated. A part of the sedges have perfect flowers (Fig. 9), but in a large number of species (*Carex*) they are unisexual, both fertile and sterile flowers occurring in the same spike or flower-cluster, or in different spikes on the same plant (Fig. 10), or more rarely on different plants. Self-fertilization is not uncommon in both families.

FIG. 9. Cotton-Grass. *Eriphorum virginicum*
A wind-pollinated sedge

THE FLOWER AND THE BEE

A few grasses bloom in the afternoon, but the majority open in the earlier part of the day, many at sunrise or a little later. Let us go out into the fields at four o'clock on a morning early in July. The sun has not yet appeared above the horizon, but a clear sky betokens a fair day. There is hardly a breath of wind, and so still is the air that

> "One might well hear the opening of a flower."

There is a legend that in ancient Egypt when the first rays of the rising sun fell on the gigantic statue of Memnon, of which only a shattered fragment now remains, there issued from it a sound which was believed to be the voice of the god. But it is a nobler greeting, the actual culmination of their life cycle, with which the flowering grasses welcome the great source of life and light.

The eastern sky has long been tinged with red, and at last the sun appears above the hills and its beams overflow the world. With the gradually rising temperature the glumes, or bracts, which protect the grass-flowers, separate and the anthers protrude, pushed out by the rapidly lengthening filaments, which grow several millimetres in a few minutes. At a quarter before five the stamens, which attain their full length in about fifteen minutes, of the common herd's-grass (*Phleum pratense*) are fully grown. (Fig. 11.) The anthers are versatile or delicately hinged at their centres so that they hang perpendicularly downward. At the lower end of each anther a narrow slit appears, which slowly extends upward, the ends of the anther becoming spoon-shaped to prevent the too rapid escape of the pollen. A slight breeze is stirring. From time to time little clouds of pollen-dust are shaken out, which descend diagonally to the ground, not infrequently without effecting pollination. In an hour's time nearly all the spikes had

32

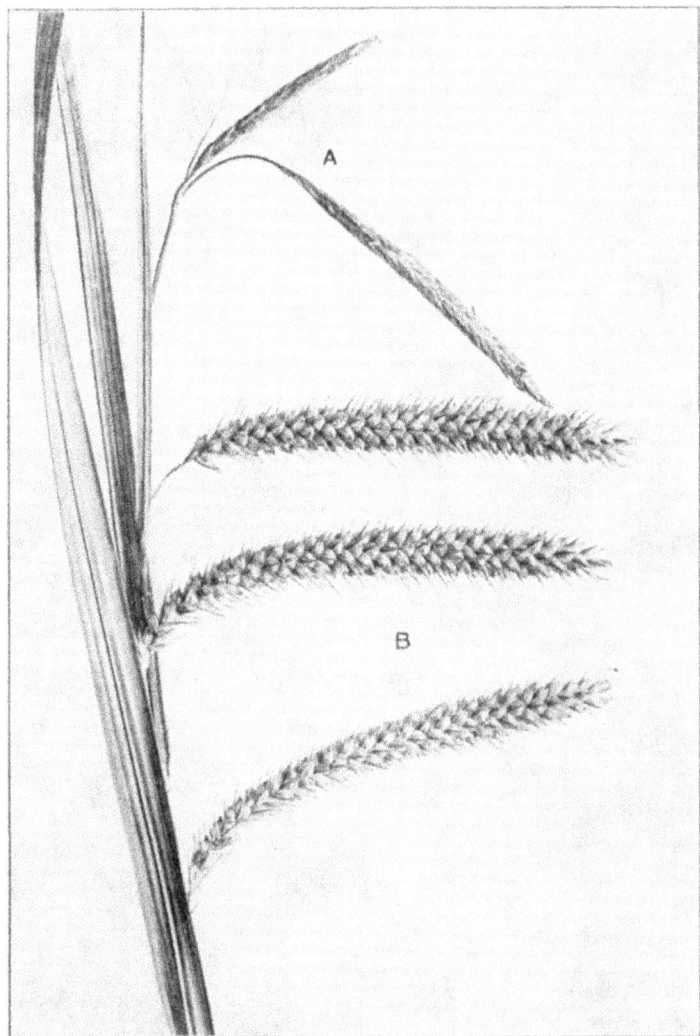

Fig. 10.   Fringed Sedge.   *Carex crinita*

A, staminate catkins; *B*, fertile catkins.   All the sedges are wind-pollinated

FIG. 11. Herd's-Grass. *Phleum pratense*
A grass pollinated at sunrise

FIG. 12. Quitch-Grass. *Agropyron repens*

A grass pollinated shortly after sunrise

dehisced, although there were a few belated flowers. The white, feathery stigmas were as beautiful as frost-crystals. Early as it was, I found a little syrphid fly eating the pollen from the anther tips; evidently it had breakfasted here on many previous mornings.

Quitch-grass, or, as it is popularly known, "witch-grass" (*Agropyron repens*), is also an early bloomer (Fig. 12), the flowers opening from six to six-thirty o'clock, about half an hour later than herd's-grass. While most grasses bloom before noon, Kerner gives a list in which he enumerates a species for nearly every hour in the day. In widely separated localities and under different climatic conditions there would seem to be considerable variation in the hour of anthesis. Long before noon the pollen of many field-varieties has been swept away, and no change takes place in the floral apparatus during the remainder of the day. It is from its appearance at midday that most persons judge of the inflorescence of the grasses; but it is well worth while to view it early in the morning, when, although not gaily hued, it will be seen to possess a delicate beauty of its own, which will greatly surprise those who behold it for the first time.

In northeastern America, north of Tennessee and east of the Rocky Mountains, I place the number of angiospermous or true flowers pollinated by the wind at about one thousand. The total number of wind-pollinated species in the world probably exceeds twelve thousand. Besides the deciduous-leaved forest-trees, and the grasses, rushes, and sedges, there are many coarse, homely weeds such as the pigweeds (Fig. 13), ragweeds, nettles, hops, pondweeds, sorrels, docks, plantains, hemp, and meadow-rue. They agree in having small, inconspicuous flowers, which are commonly odorless and nectarless, but which are produced in immense numbers. The pollen-grains are

FIG. 13. Purple Amaranth. *Amaranthus caudatus*
A wind-pollinated annual herb

round and smooth, while the stigmas are lobed or feathery to present as large a receptive surface as possible. They flourish in a great variety of situations and all present interesting phenomena worthy of careful observation. The Roman wormwood (*Ambrosia artemisiifolia*) blooms in the fall and is common everywhere in old fields and waste land. (Fig. 14.) The air is filled with the yellow pollen, which is believed to be productive of hay-fever. The plantains are midway between wind-pollination and insect-pollination. (Fig. 15.) The pleasant odor and nectar attract insects, but the smooth pollen-grains are likewise carried by the wind. The elastic stamens of the stinging nettles are in the bud, doubled back upon themselves and held under tension. When the flowers expand the filaments suddenly straighten and little puffs of pollen are forcibly projected into the air, appearing like minute explosions.

THE EVERGREEN OR CONE TREES

"Red-cedars blossom tu, though few folks know it,
An' look all dipt in sunshine like a poet."—LOWELL.

Vast forests of evergreen or coniferous trees, covering millions of acres, are found throughout the north temperate zone of both the Old and New Worlds. Large portions of Canada are densely forested with white pine and black spruce; in Siberia there are great tracts of pine, cedar, and larch; in Russia of Scotch fir, spruce, and Siberian larch; and along the southern shore of the Baltic of fir and Norway spruce. The aspect within these northern forests of conifers is dark and cold; there is little underbrush and the ground is bare or carpeted with mosses and lichens—"a solitude made more intense by dreary-voiced elements." Unlike the forests of the tropics, all kinds of animal life are scarce, and no bright-colored birds, butterflies, or flowers light up these sombre solitudes.

38

Fig. 14. Roman Wormwood. *Ambrosia artemisiifolia*

THE FLOWER AND THE BEE

Asa Gray and the older botanists often speak of "the flowers" of the conifers, but the naked seeds and the absence of a stigma, as well as a difference of opinion as to what constitutes a flower and what an inflorescence, are objections to this usage. It is better to restrict the word flower to the Angiosperms, or plants with the seeds in a closed seed-vessel a part of which is specialized to receive the pollen. The cone-trees and the tropical, fern-like cycads, which are also wind-pollinated, belong to the Gymnosperms.

There are in the world about 350 species of conifers (*Coniferales*), all of which are wind-pollinated. The cones are always unisexual, either staminate ("male") or ovulate ("female"). Both kinds usually occur on the same tree, but in the juniper and yew they are mostly on different trees. The fruiting cones are produced in positions where they are likely to be cross-pollinated; for instance, in the fir and spruce they are on the apex of the lofty spire, while the staminate cones are very abundant on the lateral branches.

When the evergreen forests bloom in May, "few folks," as the poet says, ever know it, for the neutral-colored cones do not usually contrast strongly with the green, needle-like leaves; but there are exceptions. A red pine covered with red-purple cones, half an inch in length and associated in clusters of thirty or more (Figs. 16 and 17), and a black spruce (*Picea mariana*) and balsam-fir bearing innumerable staminate cones about the size and color of a field-strawberry certainly present a strikingly handsome appearance. (Figs. 18 and 19.) The fertile cones are much less noticeable, but they are purple in the spruce and pale-green in the fir.

The quantity of pollen produced by the coniferous forests is almost beyond belief. Clouds of pollen rising from pine-trees are sometimes mistaken for columns of smoke. The falling

FIG. 15. English Plantain. *Plantago lanceolata*

A, fertile spikes; *B*, staminate spikes. A plant midway between wind-pollination and insect-pollination

Fig. 16. Red Pine. *Pinus resinosa*
Staminate cones. A wind-pollinated conifer

FIG. 17. Red Pine. *Pinus resinosa*
Ovulate cones. A wind-pollinated conifer

Fig. 18. Balsam-Fir. *Abies balsamea*
Staminate cones. A wind-pollinated conifer

of pollen-grains in such immense numbers that they tinge the
ground yellow in places has given rise to the reports of sulphur-
showers. In pine-forests the air is filled with pollen, which
slowly settles downward, powdering the foliage and branches
of the trees, the grass, and the ground. The wide dissemina-

Fig. 19. Balsam-Fir. *Abies balsamea*
Ovulate cones. A wind-pollinated conifer

tion of pine-pollen is favored by two bladder-like wings, which
greatly increases its buoyancy. Where everything is covered
with pollen it cannot fail to come directly in contact with the
ends of the ovules, in each of which there is a small orifice, the
micropyle, or little gate. This opening is filled with a drop of
mucilaginous liquid, well shown in the yew and running juniper,
to which the pollen adheres and is subsequently drawn into the
ovule by its reabsorption. Coniferous pollen is not attractive

to bees, probably because it is too resinous; but it is not infrequently eaten by beetles.

In a prolonged calm the portion of the pollen which falls from the anthers is not wholly lost, but is mostly caught and temporarily lodged on the backs of the cone-scales standing underneath, from which it is later swept away by the wind. The scales of the red pine are reflexed so that their upper surfaces form little trays, while those of the common juniper are bent backward into little pockets. These structures are apparently not adaptations but incidental results, for in the fir the apex of each scale is bent downward, the base is narrow and a concavity is almost absent. In wet weather, moreover, the little pockets are apt to gather moisture, although this is partly prevented by their expansion and consequent closure.

Wind-pollination is the oldest and most primitive method of pollen dispersion, and for millions of years the only form of pollination in existence. There were wind-pollinated trees as far back as the Devonian, and in the Mesozoic Age a remarkably equable climate prevailed over a large extent on the land surface of the globe and gymnospermous trees were the dominant forms of plant life. There were conifers, cycads, maidenhair-trees, and cycadophytes in the greatest variety, and they were all pollinated by the wind. Insect-pollination is a comparatively recent event. The efficiency of anemophily is proven not only by its age, but also by the fact that if the number of individuals is considered rather than the number of species, then far more plants are pollinated to-day by wind than by insects.

CHAPTER IV

BEES AS BUILDERS OF FLOWERS

AS pollinators of flowers, the bees, or *Anthophila* (flower-lovers) far surpass all other insects in importance. In their adaptations for collecting pollen and nectar, in diligence, and in mental attributes, bees stand easily in the first rank. To them more than to any other insects is due the evolution of our flora. Unlike all other flower-visitors they collect pollen, and it is this habit which has gained them their pre-eminence in the floral world. The beetles, flies, and butterflies take no thought for their young except to select a suitable place in which to lay their eggs. The solitary wasps provision their nests with flies, spiders, beetles, and other insects, which by stinging they have left paralyzed and helpless or dead; while the social wasps go a step further and masticate their prey before feeding it to their young. But bees are the only insects which feed their offspring with pollen; they are thus wholly dependent upon flowers, both for food for themselves and their brood.

As the result of this interdependence of bees and flowers, united with the industry and mental acuteness of the former, there has been developed a great company of bright-colored blossoms, which are especially adapted to their visits, and are, in consequence, called "bee-flowers." They agree in having the nectar deeply concealed, where it is inaccessible to ants and other pillagers. They are often irregular in form, as in the pea, bean, and snapdragon. The object, so far as it is not an incidental result, of these odd and sometimes bizarre forms

47

is to compel the bee to pursue a fixed path to the nectar so that pollination may be effected with greater certainty. Finally they are more often blue than any other color. So dependent are many flowers upon the visits of bees that, in their absence, they fail to produce seed. Such are the red clover, salvia, lark-spur, and some orchids.

Irregular or one-sided bee-flowers occur in large numbers in the violet, pea, mint, and figwort families. The nectar is usually not deeper than 7 mm., and the visitors are chiefly honey-bees, bumblebees, and long-tongued solitary bees.

The species of the violet family consist chiefly of bee-flowers, the general form of which is familiar to every one. One warm, clear day in early May I found the round-leaved yellow violet (*Viola rotundifolia*) blooming luxuriantly beneath an old beech-tree. Bumblebees, as well as smaller bees belonging to the genera *Nomada* and *Andrena*, were flitting busily about from flower to flower. On the other hand, our wild blue violets are very sparingly visited by bees or any insects, and are often infertile. (Fig. 20.) This is doubtless the reason why many species produce, besides their showy blossoms, small green flowers (cleistogamy flowers), which never expand but are very fruitful.

The beautiful and richly variegated varieties of the pansy (*Viola tricolor*) have been produced partly by selection and partly by hybridization. The corolla may be pure white, yellow, red, blue, purple, or black, or there may be manifold combinations of these hues. (Fig. 21.) These striking diversi-ties result from various mixtures and modifications of two pig-ments contained in the epidermis—violet-colored sap and yellow granules. In the pansy the spur at the base of the lower petal contains the nectar. The anthers lie close together, forming a cone, into the centre of which is shed the dry pollen;

and directly in front of this cone stands the stigma, on the lower side of which there is a flexible, lip-like projection. To reach the nectar a bee must run its tongue through the centre of the

Fɪɢ. 20. Blue Violet. *Viola cucullata*

A bee-flower; the veins on the lower petal serve as nectar-guides; the spur, or nectary, is shown in the two lower flowers

cone of anthers; and when it is withdrawn the grains of pollen cling to its moistened surface, while the lip of the stigma bends it away so that self-pollination is prevented. But when the proboscis is inserted in another flower a portion of the pollen is lodged on a receptive part of the stigma, where it soon germinates.

The pea (Fig. 22), bean, clovers, locust, vetch, alfalfa, and a host of leguminous allies, are grouped together in the pea family, or *Papilionaceæ*—a name derived from the Greek word for butterfly, because of a fancied resemblance of the flowers to that insect. Most of the 5,000 described species are bee-flowers, although there are a few bird and butterfly flowers, and at least 13 species are known which, in the absence of bees, are infertile. For the most part 9 of the 10 stamens unite to form a tube, at the bottom of which lies the nectar, if present. Four of the petals interlock around this tube, while the fifth, called the standard, is broad and erect and brilliantly colored to attract the attention of insects. In order to obtain the nectar, a bee rests upon the two lateral wing-petals, which act as levers, braces its head against the standard, when, if it is strong enough, the keel is depressed, and it may introduce its tongue in an opening at the base of the staminal tube. In this family the pollen is applied to the under-side of a bee's body in four different ways: It may escape through a valve, or it may be pumped out, or brushed out, or there may be an explosive mechanism.

But what if it is not strong enough to depress the keel? Then it does not get the nectar. Notwithstanding their beautiful hues and honey-like perfume, the flowers of the sweet pea (*Lathyrus odoratus*) are very rarely visited by bees. Why is this? So firm is the closure of the flower that only a very powerful bee is able to depress the keel and obtain the nectar. Doubtless in its transatlantic home (for the sweet pea is not endemic in America) there are bees strong enough to push open the flower. During one summer observations continued for several weeks failed to reveal a single visit by any species

FIG. 21. Pansy. *Viola tricolor*
A bee-flower; yellow flowers with dark-purple centre

of bee. But in late September, after the autumnal honey-flow from the goldenrods was over, I repeatedly saw honey-bees examine the blossoms, but they made no attempt to depress the keel. None of their efforts to find the nectar proved effectual. Neither are the bumblebees more successful. But I have seen a queen of *Bombus fervidus* visit twenty flowers in succession,and obtain the nectar illegitimately through a crevice between the standard and a wing-petal. Until the summer of 1912 I did not suppose that there was a single species of our native bees which could pollinate the flowers. But in September a female leaf-cutting bee (*Megachile latimanus*), a large and powerful insect, put in an appearance. She easily depressed the keel and, thrusting her tongue beneath the standard into the staminal tube, sucked the nectar for a long time. She was so fearless that I was able to stroke her back with my finger. As the flowers of the sweet pea are easily self-pollinated, they are very prolific, even in the absence of insects.

Another species belonging to the pea family, which is also pollinated chiefly by leaf-cutting bees, is alfalfa, the most important of fodder-plants in the West, and a most valuable source of honey to beekeepers. The flowers are known as explosive flowers. The anthers and stigma are held in the keel under tension. When a bee presses down the wings and keel, it has been described as pulling two triggers and firing off the flower, that is, the anthers and stigma fly forcibly upward and the pollen is thrown against the body of the bee. Three leaf-cutting bees (*Megachile latimanus*) have been observed to trip the flowers at the rate of 552 per hour. Honey-bees make a practice of stealing the nectar through a hole in the side of the corolla, but are of little benefit as pollinators. East of the Mississippi alfalfa yields very little nectar, but in the Western

FIG. 22. Garden-Pea. *Pisum sativum*

A bee-flower, probably from Asia; in America it is usually self-pollinated

FIG. 23. Alfalfa. *Medicago sativa*

Cultivated for fodder, yields nectar in the West, but not in the East. A bee-flower, pollinated chiefly by leaf-cutting bees and bumblebees

Fig. 24. White Clover. *Trifolium repens*

First stage; all the flowers point upward. The most important honey-plant in North America

States hundreds of tons of honey are stored from it annually. (Fig. 23.)

Many of the flowers of the pea family, in fading, undergo a change in position and color. In the white clover the white central flowers contrast with an older outer ring of rose-colored blossoms. (Figs. 24 and 25.) In the yellow clover the newer flowers contrast with a ring of chestnut-brown. In the wild vetch (*Vicia Cracca*) the older flowers bend downward and turn from violet-blue to dark purple (Fig. 26), while the purple flowers of *Desmodium* become green in withering. In this way the bees are able to distinguish the older, nectarless flowers which have been pollinated from those which have just opened.

In the mint family (*Labiatæ*) and figwort family (*Scrophulariaceæ*) the flowers stand horizontal and are more or less two-lipped. According to the way in which the lips have developed the larger flowers exhibit strange, fantastic forms which mimic the heads of reptiles or other animals, or of inanimate objects. Such are the turtlehead, snapdragon, monkey-flower (Fig. 27), toadflax, foxglove, skullcap, and *Calceolaria*, or shoe-flower. (Fig. 28.) A great many species are pollinated by bees, as the hedge-nettle (*Stachys*), dead-nettle (*Lamium*), hemp-nettle (*Galeopsis*), the sages, cow-wheat (*Melampyrum*), and the loose-wort (*Pedicularis*). (Figs. 29 and 30.) A goodly number are bumblebee-flowers, some exotic species are bird-flowers, the figworts (*Scrophularia*) are wasp-flowers, while others with small, inconspicuously-lipped corollas are fly-flowers, as the speedwells and mints.

It will be remembered that in the pea family by means of various devices the pollen was placed on the under-side of the bee's body; but in the mints and figworts it is usually deposited on the back of the insect. In the turtle-head (*Chelone glabra*) the four heart-shaped anthers lie well forward in the angle

FIG. 25. White Clover. *Trifolium repens*

Second stage; after pollination the flowers bend downward and turn reddish, then brown

FIG. 26. Purple Vetch. *Vicia Cracca*
A bee-flower, common in fields

formed by the sides of the upper lip. (Fig. 41.) Their inner faces are applied together to form a single chamber or cavity into which the pollen, when ripe, falls. The contiguous edges

FIG. 27. Two varieties of Monkey-Flower. *Mimulus luteus*
A bee-flower

are densely woolly to protect the pollen and to prevent the relative displacement of the anthers by connecting the first pair above, and the first and second pairs at the sides. When

59

a bumblebee enters the flower it spreads apart the arched filaments, opening the pollen receptacle, and covering the thorax with fine grains of pollen.

Since both methods of lodging the pollen on the body of a bee have proved effective, it may be inquired why have two directly opposite courses been followed in different families. To answer this question it is necessary to go back to an early stage in the history of the pea, mint, and figwort families, when as yet the flowers were wheel-shaped or regular in form. In the ancestral plants from which the pea family was derived the stamens were long and projected far out of the corolla, so that insects alighted on the anthers and filaments, pushing them downward. But in the primitive stock of the mint and figwort families the stamens were short and nearly included in the corolla-tube. When a bee rested on the lower lip its head came beneath the anthers, which were pushed up against the helmet, just the opposite of what happened in the pea family. The pollen was consequently placed on the bee's back. If the anthers stood directly within the entrance to the corolla, as in the violet, then the bee was compelled to run its tongue between them. Thus in the evolution of these families it was the length of the stamens which determined where the pollen should be placed on the insect visitor.

Regular flowers, like the buttercup and rose, always stand vertical, that is, if erect they face the sky, or if pendulous the earth. Irregular, or bilaterally symmetrical flowers, on the contrary, always stand horizontal, or face the horizon. This is well shown in the dense flower-cluster of the horse-chestnut, where the lateral flowers are irregular and the single terminal flower rotate or regular. Vertical flowers, like the borage or strawberry, are approached by insects with equal ease from every side, and the forces which might change their

FIG. 28. Yellow Calceolaria. *Calceolaria scabiosæfolia*

A bee-flower; the corolla is two-lipped, with the lower lip much the larger, sac-shaped

form are thus held in equilibrium or counterbalance each other. But where the flower stands horizontal, as the snap-dragon or the sage, bees almost invariably land on the lower

Fig. 29.   Yellow Rattle.   *Rhinanthus Crista-galli*
A bee-flower

side of the corolla.   As a consequence one, two, or three of the lower petals become transformed into a lip which serves as a landing-stage.   Such an adaptation is beneficial to both the flowers and the guests. Contemporaneously the upper petals were modified into a helmet, or galea, to protect the anthers and the interior of the flower from wet—and a flower

FIG. 30. Heal-All. *Prunella vulgaris*
A violet-colored bee-flower belonging to the mint family

like the snapdragon is evolved. In no two cases were the conditions exactly alike, with the result that a great variety of forms has been produced, each of which has its special explanation. Some are more perfect than others, while some have retrograded and lost their earlier structure. But the general principle has been the same, although it has been worked out in manifold ways. (Figs. 31 and 32.)

> "We are groping here to find
> What the Thought which underlies
> Nature's masking and disguise."

But all bee-flowers are not one-sided (zygomorphous) or irregular. They may be funnel-formed as in the gentians, or urn-shaped as in the checkerberry and blueberry, or even wheel-shaped as in the common borage, or regularly spurred as in the pendulous flowers of the columbine. (Fig. 38.)

It is a remarkable fact that bee-flowers are more often blue than any other color. Let us look at the colors of these flowers in the Northern States. Of 34 species of violets 17 are blue, 4 blue-purple, 6 yellow, and 7 white. Of 197 species belonging to the pea family (*Papilionaceæ*) 24 are blue, 88 blue-purple, 13 red, 33 yellow, and 39 white. Of 120 species of the mint family (*Labiatæ*) 33 are blue, 47 blue-purple, 12 red, 4 yellow, and 24 white. Of the 113 species of the figwort family (*Scrophulariaceæ*) 28 are blue, 32 blue-purple, 7 red, 33 yellow, and 13 white.

On the other hand, neither bee nor blue flowers occur in the pink and rose families. In the immense orchis family, in which bee-flowers are of comparatively rare occurrence, there is only one blue flower, *Vanda cærulea*, from India. In this family red is developed much more easily than blue. It should be added that, when two or more species of bee-flowers belonging to the same genus blossom simultaneously in the same locality, they are frequently unlike in color, as the red, white, and yellow clovers. This diversity of color facilitates the efforts of the bee to remain constant to one species.

If you inquire why bee-flowers are so often blue, I shall be compelled to admit that I do not know with certainty. It is a problem which still awaits further study. Some naturalists have said that bees prefer blue to every other color, while others claim that it is merely an incidental result correlated with the higher specialization of the flower. For example, in the animal kingdom, white cats (if they have blue eyes) are nearly always deaf, but no one knows why.

Bee-flowers are usually marked with spots or lines called "nectar-guides," which point out the way to the nectar. (Fig. 34.) In the snapdragon the palate is yellow; in the pickerel-weed there are two bright-yellow spots on the middle lobe of the upper lip; in the turtle-head the white corolla has reddish lips. The flower of the hedge-nettle (*Stachys erecta*) is yellowish white, with the border of the upper lip marked with two purple stripes and the lower lip purple-spotted. The flower

Fig. 31. Pink-Fringed Polygala. *Polygala paucifolia*

A bee-flower. The crest, or fringe, is well shown in the photograph

FIG. 32. Pink Lady's-Slipper. *Cypripedium acaule*

This flower is pollinated by small bees, which gain an entrance at the front end of the slipper
and pass out through two small holes at its base

FIG. 33. Sheep-Laurel. *Kalmia angustifolia*

A regular or wheel-shaped bee-flower. The stamens are elastic and when touched by the legs of a bee the anthers, which are held in little pockets in the corolla, are released and fly upward, throwing the pollen over the bee

of the dead-nettle (*Lamium album*) is large, white or sometimes rose-colored, with the under lip pale yellow, marked with olive-colored dots; while the flowers of the hemp-nettle (*Galeopsis*

FIG. 34.   Foxglove.   *Digitalis purpurea*
A bumblebee-flower, with the corolla spotted on the lower inner side

Tetrahit), so common in waste places, is purple, with a path-finder on the lower lip of a yellow spot crossed by a network of red lines.

The marvellous adaptations of flowers for effecting pollination, both by their variety and ingenuity, fill us with astonishment, and occasionally they surpass the bounds of the wildest imagination.   The opening, maturity, and fading of the flower, the various movements of its organs, the allurements of color, odor, and nectar, and the behavior of the insect guests, which may number from one to more than three hundred, afford an endless field for observation. Flowers cease to be merely bright bits of color in the landscape when we know their life histories, their rivalries, and tragedies; and—yes, their comedies—we see as upon the stage reflections of our own experiences. There is no more fascinating study than entering the secret chambers of these bright-hued floral edifices which adorn our fields and gardens and probing the mysteries which there confront us. But we should seek the living blossoms.

"Each one of the beautiful flower faces," says Hermann Mueller, "which we were wont to marvel at with a sad feeling of resignation as so many mysteries forever veiled now looks upon us, inspiring hope, and stimulates us in friendly wise to cheerful perseverance, as if it would say, 'Only venture to come to me, and in true love make yourself acquainted with me and all my conditions of life, as intimately as you may, and I am ready to let fall the veil that hides me, and trust myself and all my secrets to you.'"

"Think of all these treasures,
Matchless works and pleasures,
Every one a marvel, more than thought can say;
Then think in what bright showers
We thicken fields and bowers,
And with what heaps of sweetness half wanton May.
Think of the mossy forests
By the bee-birds haunted,
And all those Amazonian plains, lone lying as enchanted."

CHAPTER V

BUMBLEBEE–FLOWERS

THE English nation owes its power and wealth largely to bumblebees. This statement sounds a bit sensational, not to say improbable. But it was the opinion of a distinguished German scientist, Carl Vogt, and is indorsed by an eminent living biologist, Ernst Haeckel, of Jena. Let us examine the evidence.

Red clover (*Trifolium pratense*) is chiefly pollinated by bumblebees, and is, therefore, called a bumblebee-flower. (Fig. 35.) In the absence of bumblebees this valuable and extensively cultivated fodder-plant yields little or no seed, as may be easily proven by covering a small bed of red clover with netting to exclude bees. According to Carl Vogt, one of the most important foundations of the wealth of England is found in the cattle, which feed principally on red clover. "Englishmen," says Haeckel, "preserve their bodily and mental powers chiefly by making excellent meat—roast beef and beefsteak—their principal food. The English owe the superiority of their brains and minds over other nations to their excellent meat." There is no need to enter here into any discussion as to whether Haeckel's logic is sound or not, but it will be hardly denied that the production of this meat depends to a great extent on the industrious bumblebee.

Incidentally it may be added that Darwin pointed out that the number of bumblebees in England was determined by the number of cats. Mice rob bumblebees' nests and are in turn killed by cats; consequently if there are few cats there are many mice and few bumblebees. Here Huxley suggested that,

70

as old maids are fond of cats, and usually keep one or more of these animals as pets, it depended on them whether there should be an abundant crop of red clover or not. Let us, then, chivalrously admit that, as in the bee-hive or bumblebee's nest, the existence of the colony depends on the workers, or unmated females, so the prosperity of England depends on her old maids. It is certainly a curious instance of the intimate correlation of every part of nature.

When the farmers of New Zealand attempted to grow red clover in their fields they learned to their cost its dependence on bumblebees for pollination, for it failed to produce seed. On inquiry it was learned that there were no bumblebees in these large islands, and it was not until after the introduction of several species from England that the raising of clover-seed became commercially profitable. Once introduced, the bumblebees multiplied apace; and a few years ago a letter appeared in one of the New Zealand papers complaining that they were becoming so numerous that they threatened to consume the nectar of all the flowers and leave none for the domestic bee. But the alarm proved groundless, for in 1905 the Canadian Department of Agriculture received a letter from the secretary of an agricultural association in New Zealand inquiring what species of *Bombus* pollinated the red clover in that country. Three species of bumblebees (*Bombus terrestris*, *B. hortorum*, and *B. hortorum* variety *harrisellus*), descended from those imported in 1885, are stated to occur in New Zealand; but *B. terrestris*, the most abundant species, was regarded as unsuitable for clover pollination owing to the shortness of its tongue. It was believed that the best results had not yet been obtained, and that it was desirable to introduce longer-tongued species. Of American species *Bombus americanorum* and *B. fervidus* appear well adapted for this purpose.

THE FLOWER AND THE BEE

To-day bumblebees benefit the islands of New Zealand annually to the extent of many thousand dollars. Fields which were almost barren in their absence now produce great quantities of seed. At Canterbury 26 acres of red clover were the resort of thousands of bumblebees, and yielded 400 to 500 pounds of seed per acre. In one province alone, in 1912, 610 acres were sown with red clover, which it is estimated yielded an average of 158 pounds to the acre.

The nectar of red clover is secreted at the base of a floral tube a little over 9 mm. long, where it is beyond the reach of honey-bees, which have a tongue only about 6 mm. in length. Under normal conditions, then, honey-bees do not resort to the red-clover fields; but occasionally in very dry weather the floral tubes become so short that large yields of honey are obtained. Two or three times during the past thirty years at Borodino, N. Y., red clover has been a very valuable source of honey; and one season full 60 pounds, on an average, to a colony was obtained. A very remarkable illustration of the relation of rainfall to the length of the corolla-tubes of red clover was observed by an apiarist at Medina, Ohio. One of his apiaries was located near Medina, and another about two miles north of that city. A few years ago there was a drought at the north bee-yard, and the floral tubes of the red clover were so much shorter than usual that honey-bees were able to reach the nectar. When one of the farmers began to cut his field of red clover that season, the cutter knives of the mower stirred up so many bees that they attacked the horses and their driver. Singularly enough at Medina and the south bee-yard there was an abundance of rain. The red clover made a luxuriant growth, and the floral tubes were so long that the bees could not obtain the nectar, and consequently there were none on the clover-heads. Thus two bee-keepers, living only a few miles apart,

Fig. 35. Red Clover. *Trifolium pratense*
A bumblebee-flower

might have arrived at diametrically opposite conclusions as to the value of red clover as a honey-plant.

Three other genera of very common bumblebee-flowers may be found in almost any old-fashioned garden. They are the larkspurs (*Delphinium*), the aconites, or monk's-hoods (*Aconitum*), and the columbines (*Aquilegia*). They all agree in having the nectar concealed in long spurs or nectaries, which vary in length in the different species. The tongues of the various species of bumblebees also differ in length, ranging in the workers from $\frac{5}{16}$ to $\frac{10}{16}$ of an inch. In the females, or queens, the tongue is still longer, and in the garden-bumblebee of Europe reaches the length of $\frac{13}{16}$ of an inch.

Of our hardy perennials there are few which produce a more stately effect than the bee-larkspur (*Delphinium elatum*) with its wand-like racemes of deep-blue flowers. This plant is a native of Europe, where it is pollinated by the female of the garden-bumblebee, no other bee on the wing at the time it blooms having a tongue long enough to reach all of the nectar, although a part of it is accessible to a few other bees. I have seen honey-bees searching the flowers and pushing their tongues down into the long spurs as far as possible, but they were never able to gain any of the sweet spoil. (Fig. 36.)

The aconites, or, as they are perhaps better known, the monk's-hoods, are, says Kronfeld, bumblebee-flowers *par excellence*. When a plaster-cast is made of the inside of a flower it is found to correspond almost exactly to the shape of a medium-sized bumblebee. This genus of plants is so dependent on bumblebees for pollination that it is absent from those parts of the world where there are no bumblebees. For instance, there are no native bumblebees in Australia, Arabia, South Africa, and New Zealand, and in these countries there are no indigenous species of *Aconitum*. (Fig. 37.)

74

FIG. 36. Bee-Larkspur. *Delphinium elatum*
A bumblebee-flower

The columbines manage to thrive and bloom under the most difficult conditions of soil and climate. The long spurs of the variously colored, pendulous flowers of the garden-columbine

FIG. 37. Monk's-Hood. *Aconitum Napellus*
A bumblebee-flower

(*Aquilegia vulgaris*) are rich in nectar and are great favorites of bumblebees. (Fig. 38.) Sometimes they bite holes in the spurs in order to save time, and then the honey-bees also come and suck the nectar through these punctures. Our wild colum-bine (*A. canadensis*) has scarlet flowers which are yellow inside, or rarely all over, and is chiefly visited by humming-birds.

Other native humming-bird flowers are the cardinal-flower, the trumpet-honeysuckle, the painted cup, and the trumpet-flower; but bird-flowers are not common in North America, although abundant in tropical South America.

Another common bumblebee-flower is the garden-nasturtium.

Fig. 38.   White Garden-Columbine.   *Aquilegia vulgaris*
A bumblebee-flower

The lower part of each petal is marked with red, which serves as a guide to the nectar; while the claws of the lower petals are fringed with hairs which prevent water from running into the spur. Honey-bees cannot reach the nectar, although they occasionally attempt to do so, for only bumblebees with the

77

longest tongues can obtain all of it. The spur is so tough that it cannot be perforated. Honey-bees gather pollen from the anthers, which open one at a time rising successively before the mouth of the flower.

The snapdragon (*Antirrhinum majus*) is another bumblebee-flower widely cultivated in gardens. So firmly are the lips closed together that the smaller bees cannot force them apart, and thus the nectar is protected for the rightful guests. But as the flowers grow older the lips part slightly, and then the smaller bees are able to force an entrance. The great size of the corolla permits the largest bumblebees to creep wholly within it. (Fig. 40.)

A typical wild bumblebee-flower is the turtle-head (*Chelone glabra*), which grows along the banks of streams and in marshes. The large, white flowers rudely mimic in form the head of a turtle. Although I have had them under observation for many hours and on many different occasions I have never seen them entered by any insects except bumblebees. Wasps and flies sometimes examine the lips, which are tinged with yellow, apparently looking for nectar; but they never pass between them into the corolla-chamber. The mouth of the flower is so small that a bumblebee sometimes finds difficulty in entering, but once inside there is an abundance of room for a bee to turn completely around. I once placed several flower-clusters of the turtle-head in a glass of water a few feet in front of a bee-hive; but of the many honey-bees constantly coming and going not one of them entered a flower. But presently, notwithstanding their unusual position, every blossom was examined by bumblebees. The honey-bees seemed instinctively to know that these flowers were not designed for their use. (Fig. 41.)

The common "touch-me-not," or jewelweed (*Impatiens biflora*) which covers acres of damp land, is another bumble-

Fig. 39. Tartarian Honeysuckle. *Lonicera tartarica*
A bumblebee-flower

bee-flower, much sought after by *Bombus vagans* and *B. terricola*. Its brown-spotted, orange blossoms are shaped like a horn of plenty with the spur inflexed or bent inward beneath it. The flower is suspended horizontally, with the anthers and stigma lying on its upper side, so that when a bee enters the dilated corolla-sac its back is dusted with pollen which is carried away to another flower. It is a matter of some difficulty and delay for bumblebees to enter the flowers, and very likely the short-tongued workers are not able to reach all of the nectar; so after a little while they bite holes in the spurs and steal the sweet contents. On August 10 I examined a large number of flowers, but none of the nectaries were punctured and they were visited normally by *Bombus vagans*, or the wandering bumblebee, at the rate of 7 to 12 visits per minute. But during the latter part of August I found hundreds of the spurs perforated and both bumblebees and honey-bees gathering the nectar from these punctures. (This habit led Mueller to call the bumblebee an "anti-teleologist.") A honey-bee was watched during 25 successive visits, and not once did it even make a pretense of entering the flower; but in every instance it swung itself astride of the spur, pushed its tongue through the puncture and became literally a flower-robber. Ten such visits may be made in a minute. (Fig. 42.)

If after the manner of plants famous in myth and story the *Impatiens* (fitly called "touch-me-not" in this respect) could speak, what a protest it would utter! For unknown centuries this floral edifice has been under construction, only at the last to have its usefulness threatened by a change in the habits of its bee visitors. Humming-birds also visit the flower, while small beetles and spiders occasionally seek shelter in the sac.

But not all bumblebee-flowers are irregularly shaped. The closed gentian and the fringed gentian, both of which are pol-

FIG. 40. Snapdragon. *Antirrhinum majus*
A bumblebee-flower

Fig. 41. Turtle-Head. *Chelone glabra*
A bumblebee-flower

linated by bumblebees, are funnel-formed. The gentians bloom at the close of autumn and are very abundant in the Alps, where they display broad expanses of blue color. The closed

Fig. 42. Jewelweed. *Impatiens biflora*
A bumblebee-flower

gentian never opens, and on a cold morning the temperature within the corolla-chamber is often several degrees above that of the outside atmosphere. The fringe on the edge of the corolla of the fringed gentian prevents the ingress of small injurious insects. (Fig. 43.)

Other familiar bumblebee-flowers are the beautiful *Rhodora canadensis*, which is pollinated in spring by queen bumblebees, the only caste of bumblebees then on the wing, for the males and workers do not appear until later (Figs. 44 and 45); the fly-honeysuckle (*Lonicera ciliata*), also pollinated in May by female bumblebees, which in their haste to get the nectar often cut the buds into shreds; the Tartarian honeysuckle (*Lonicera Tartarica*) of the garden (Fig. 39); the bog fly-honeysuckle (*Lonicera cærulea*); the bush-honeysuckle (*Diervilla trifida*), the yellow flowers of which turn red in fading; the horse-chestnut; the foxglove; and the *Gladiolus*.

The garden-bean is largely self-fertilized, but bumblebees visit the flowers more or less; the scarlet runner is also a bumblebee-flower, although honey-bees are often able to gather a little of the nectar. The lungwort (*Pulmonaria officinalis*), belladonna (*Atropa belladonna*), the bearberry (*Arctostaphylos Uva-ursi*), the wood-betony (*Pedicularis sylvatica*), gill-over-the-ground (*Glechoma hederacea*), and largely butter-and-eggs (*Linaria vulgaris*) are bumblebee-flowers. The scarlet sage (*Salvia pratensis*), with its walking-beam mechanism for placing the pollen on a bee's back, the dragon's-head (*Dracocephalum*, 3 species), Molucca balm (*Moluccella lævis*), bugle (*Ajuga reptans*), and several orchids, as the showy orchis (*Orchis spectabilis*), the pink flowers of *Pogonia ophioglossoides* common in bogs, and *Calypso borealis* are all pollinated by bumblebees. The pretty flowers of the purple Gerardia (*Gerardia purpurea*) are abundant in autumn, but they contain little nectar and few bumblebees visit them. Finally there may be added to the list *Cerinthe alpina*, *Scopolia atropoides*, and black henbane (*Hyoscyamus niger*).

No matter how bizarre or grotesque a bumblebee-flower may be to-day, it is derived from a primitive form that was per-

84

FIG. 43. Fringed Gentian. *Gentiana crinita*

A bumblebee-flower

FIG. 44. A Common Bumblebee. *Bombus impatiens*
1, Male; 2, queen; 3, worker

FIG. 45. American Bumblebee. *Bombus americanorum*
1, Queen; 2, worker

fectly regular, to which it may occasionally revert. The columbine sometimes produces five flat, instead of five spurred petals,
and there is a stellate form in cultivation. Even the larkspurs
and monk's-hoods may become perfectly regular. A regular
form of the snapdragon was cultivated at one time and called
"the wonder." Darwin crossed the regular forms with their
own pollen and raised a whole bed of similar flowers. In
Linaria, or butter-and-eggs, reversion to a radiate form often
occurs in many species. Even the grotesque orchids untwist
themselves and display regular star-shaped flowers, which is
the normal form in *Dendrobium normale*. Any unsymmetrical
flower may at times become symmetrical, that is, in the history
of the evolution of flowers radiate, or star-shaped, flowers are
more primitive than one-sided or bisymmetrical flowers.

Notwithstanding the industry and immense numbers of the
honey-bee, there are no flowers adapted to this species alone.
It is impossible not to inquire why there should be bumblebee-
flowers, but no honey-bee flowers. Should we not rather
expect the reverse? But on inquiry into the economy of the
honey-bee the reason is evident enough. The domestic bee
requires large quantities of stores, and in order to obtain them
it must visit a great variety of flowers throughout the entire
season. For this purpose a tongue of medium length is far
more useful than a longer one. If the tongue is very long the
nectar in open, wheel-shaped flowers like the strawberry and
basswood can be sucked up only with difficulty and delay.
To be sure, they would be able to obtain some nectar now inaccessible, as from bumblebee-flowers like the red clover, or
from moth-flowers like the climbing honeysuckle, but this
would not compensate for the disadvantages sustained. If a
longer tongue would have been beneficial to the honey-bee,
Nature would have long since developed one. Apiarists can-

not improve on the tongue of the honey-bee, nor can they produce a permanent strain of red-clover bees. Taken as a whole, and under all conditions, the tongue of the honey-bee, as it exists, is much better adapted for the work to be done than any that can be produced by artificial selection. But there would seem to be no reason why a variety of red clover with shorter corolla-tubes should not be obtained.

CHAPTER VI

THE GATHERING OF THE NECTAR

URING the honey-flow from white clover, basswood, alfalfa, sage, goldenrod, or any other honey-plant which yields nectar copiously the most intense excitement and activity prevails in the apiary. Work begins early in the morning and continues until late in the afternoon. The air is filled with thousands of bees rushing to and from the fields, and the roar of their wings may be heard at a distance from the hives. Oblivious to everything else, they are obsessed with the single purpose of garnering the golden store; and so diligently do they labor that the life of a worker bee during the summer is only forty days, whereas in winter they may live for six months or more. In a colony of 50,000 bees it has been estimated that there are 30,000 field-bees, and if each fielder makes ten trips a day then there would be a total of 300,000 visits to flowers in a single day. About 37,000 loads of nectar are required for the production of a pound of honey, and, according to the locality, a hive may gain from 1 to 10 pounds of honey in a day. It is clear that even a very slight saving of time or labor becomes in the aggregate of great importance to the colony.

It is the diligence and skill of bees—honey-bees, bumblebees, and solitary bees—in visiting flowers which makes them the most valuable of pollinators. They learn quickly from observation and are subsequently guided by the memory of past experience. Buckwheat secretes nectar freely during the forenoon and attracts thousands of bees; but during the afternoon the flow entirely ceases and the bees promptly discontinue their visits. "In spite of the shimmering sea of flowers, in spite of the strong fragrance, only a few bees can be found in the buckwheat-field after twelve o'clock." Again, a sudden shower followed by a fall in temperature may bring the buckwheat harvest to an abrupt and premature close in August, when ordinarily it would continue into September. The bees then immediately cease visiting the flowers and in countless numbers attempt to rob each other; the time of their visits thus always coincides with the period of active secretion of nectar. (Fig. 46.)

The rapidity with which bees visit flowers is greatly influenced by their form and arrangement. Honey-bees cannot reach the nectar of the yellow and red garden-nasturtiums, which lies at the bottom of a long calycine spur and, consequently, are seldom seen on the flowers, although occasionally they come for pollen. One of the larger bumblebees (*Bombus fervidus*), which has a tongue 12 mm. long, on the contrary devotes itself exclusively to sucking nectar and ignores the pollen. It is rather clumsy in its movements and visits only from 12 to 14 flowers per minute. The bilabiate flowers of the pickerel-weed (*Pontederia cordata*) are examined much more rapidly by a smaller species of bumblebee (*Bombus vagans*). In July the violet-blue spikes of this aquatic plant fringe the banks of many northern streams in countless numbers. *Bombus vagans* is a very common visitor, beginning always with the lowest flowers of the spike and working upward. By actual count, several times repeated, I found that the average number of visits per minute was about 70. The small florets of the goldenrods are visited so rapidly that the number per minute cannot usually be counted. But when the nectar is very abundant, as in the flowers of the basswood, century-plant, spider-plant (*Cleome*

FIG. 46. Buckwheat. *Fagopyrum esculentum*

Secretes nectar only during the forenoon; yields a dark honey, about the color of molasses

spinosa), and some species of *Eucalyptus*, a honey-bee may obtain a load from 2 or 3, or even 1 flower.

Bees in collecting pollen and nectar are faithful as a rule to a single species of flower—they exhibit "flower fidelity." This is for their advantage since, if they were constantly passing from flowers of one form to those of another, much time would be lost in locating the nectar. At the same time the flowers are cross-pollinated and a waste of pollen is prevented. Even whole colonies may be true to a single species. At Ventura, Cal., in 1884, 1 colony out of 200 gathered exclusively from an abundance of mustard-bloom, while 199 gathered from the sages.

But where there are several differently colored varieties of the same species, honey-bees soon learn to visit them indiscriminately. *Zinnia elegans* displays white, yellow, red, and purple varieties; *Dahlia variabilis* white, yellow, orange, red, and purple; and *Centaurea Cyanus* (bachelor's-button) red, white, blue, and purple. Bees pass freely, in visiting these flowers, from one color to another. It is obvious that the varieties differ in color alone, and are alike in form, odor, and nectar. Under these conditions bees quickly learn that it is for their advantage to ignore differences in hue.

But the flowers of many closely allied species are so similar that they puzzle even the highest authorities in taxonomy; and Asa Gray writes in one of his letters that the asters threatened to reduce him to blank despair. In such cases honey-bees cease to adhere strictly to a single species, and visit indiscriminately the different kinds of buttercups, spiræas, and goldenrods. I have also often seen bumblebees pass from one species of goldenrod to another, and even back and forth between goldenrods and asters. Occasionally I have seen them pass between very different forms of flowers, as from the sunflower

FIG. 47. Bushy Goldenrod. *Solidago graminifolia*

to the scarlet runner, or from the goldenrod to the purple vervain (*Verbena hastata*).

On the other hand, the honey-bee often displays a remarkable power of distinguishing between closely allied species even when they are of the same color. One of the common goldenrods (*Solidago graminifolia*) has its heads or capitula arranged in crowded, flat-topped corymbs. (Fig. 47.) Another common variety (*S. rugosa*) has the inflorescence panicled. (Fig. 48.) In an upland pasture these two species were found growing together, the panicled form being much the more abundant. Honey-bees, the only insects present, showed a marked preference for *S. graminifolia*, although occasionally they passed over to the other species. They were repeatedly seen to leave *S. graminifolia*, and after flying about, but not resting on the flowers of *S. rugosa*, return to the plants they had left only a few moments before. In another instance a bee was seen to wind its way among the plants of the latter species until it found an isolated plant of *S. graminifolia*. A plant of each of the above species was bent over so that the blossoms were intermingled, appearing as a single cluster; a honey-bee rested on *S. graminifolia*, and it seemed very probable that it would pass over to the flowers of *S. rugosa*, but such was not the case, for presently it flew away to another plant of the former. The behavior of these bees in their endeavors to adhere to a single species was thus attended both by loss of time and repeated visits to the same blossoms.

On another occasion the whitish or cream-colored inflorescence of *Solidago bicolor*, our one non-yellow species of goldenrod (Fig. 49), was observed to be very frequently visited by the males of *Bombus ternarius*, while the yellow-flowered goldenrods in the vicinity were entirely neglected. By holding yellow-flowered clusters directly in their way I repeatedly induced the bumblebees to leave *S. bicolor;* but they quickly perceived that they had passed to a different flower, and in-

FIG. 48.   Tall, Hairy Goldenrod.   *Solidago rugosa*
Yields a heavy golden-yellow honey

variably after a few seconds or sometimes instantly returned to the cream-colored species.   They were probably influenced by the greater supply of nectar in the flowers of *S. bicolor*, for the plants were growing on burned land and were of larger size than usual.

Butterflies are much less particular in their visits, and I have frequently seen the silver-winged butterfly (*Argynnis aphrodite*) fly back and forth between the flowers of the common elecampane (*Inula Helenium*) and the Canada thistle. This yellowish-red butterfly was flitting about upon the large yellow flower-heads of *Inula*, for which it showed a decided preference to the purple flowers of the thistle. The white cabbage-butterfly, on the contrary, which was common on the thistle-bloom, confined its visits chiefly to that species. This singular behavior must have been determined by other causes than the color of the flowers. In illustration of the irregular habits of flies in visiting flowers, I may mention that I have seen the syrphid fly *Mesograpta germinata* pass from the water-horehound (*Lycopus europœus*) to the tear-thumb (*Polygonum sagittatum*) and thence to the blossoms of the smaller willow-herb (*Epilobium molle*), where its career ended in my collecting-net.

When the nectar is deeply concealed in irregular flowers or long nectaries, as in the larkspur, clover, columbine, fly-honeysuckle, and skullcap, a much greater amount of time is required to gather it than when it is fully exposed. Such flowers are chiefly adapted to the skilful, long-tongued bees, while beetles, flies, wasps, and many other insects are either unable to find it or have a tongue too short to reach it. In order to obtain the nectar more easily than by entering the flowers in the legitimate way certain bumblebees have formed the curious habit of biting holes in the nectaries or corolla-tubes. The holes are made by the laciniæ or lance-shaped ends of the maxillæ. The maxillæ, which are the second pair of jaws and are situated just below the mandibles, are composed of two joints, a basal part called the stipe, and a terminal acutely pointed blade or

Fig. 49. Cream-Colored Goldenrod. *Solidago bicolor*

lacinia. When these two sharp points are moved back and forth on the outer side of a nectary they may puncture it, making either a single slit or two small holes side by side.

The fly-honeysuckle (*Lonicera ciliata*) is a graceful, slender shrub, which blooms in northern woodlands during the last weeks in May. The flower-stalk bears at its summit two pendulous, yellowish-green flowers, which are tubular and half an inch in length. The nectar is secreted and lodged at the base of this tube, where it can be reached by the long tongues of bumblebees, by which the flowers are pollinated. The female of *Bombus vagans* was often observed stealing the sweet secretions through holes in the buds. Sometimes the perforation was near the apex, but usually it was near the base of the tube, and in one instance I found the corolla nearly circumcised and held only by a few threads. (Fig. 50.)

Bumblebees also puncture at the apex (usually on the underside) the buds of the common skullcap (*Scutellaria galericulata*), even when they are quite immature. The flowers are bilabiate, or two-lipped. In two instances I observed a narrow slit on the under-side of the corolla-tube, and in a third case the whole upper portion of the tube was cut away, leaving the lips suspended by a mere thread. Hundreds of spurs of the wild balsam (*Impatiens biflora*) are perforated on the under-side; sometimes there are several holes, in other cases a single slit. After the punctures are once made honey-bees rob the flowers as well as bumblebees, making about ten visits per minute.

The garden-columbines secrete nectar very plentifully. If a flower of the white variety be held so that the light shines through its translucent tissue, the nectar may be seen filling a tenth of an inch of the hollow spurs or nectaries. Both the purple and white varieties are punctured by bumblebees. Mueller observed a bumblebee, after a fruitless endeavor to obtain the nectar, bite a hole in the spur; and afterward it punctured the flowers visited without any preliminary delay. I have noticed three distinct incisions, one above the other, on a petal of this plant. The first was over half an inch from

FIG. 50. Fly-Honeysuckle. *Lonicera ciliata*
A bumblebee-flower; in their haste to obtain the nectar bumblebees often puncture the corolla

the tip of the spur, well up on the expanded part of the tube; the second was lower down, and the third still nearer the tip. Apparently the upper puncture was too far distant to permit the tongue of the bee to reach the nectar, and to rectify this mistake the other holes were made lower down.

THE FLOWER AND THE BEE

Although honey-bees freely rob the nectaries after they have been punctured by bumblebees, they are probably not able themselves to bite holes in them. On August 14, 1909, the vines of the scarlet runner in my garden were a blaze of glory. Honey-bees and bumblebees were constantly coming and going, but not one of them visited the flowers in the normal way. There was a hole on the under-side of every nectary; and the bees went directly to these holes, out of which they easily sucked the nectar. The punctures were all on the left-hand side, which may be explained by the fact that the larger bees almost invariably alight on the left wing, for the reason that the spirally coiled carina closes the entrance beneath the standard on the right-hand side.

In the spring of the following year, 1910, I planted 5 hills of the scarlet runner bean at a distance of about 50 feet from my apiary. By the last of July it was in bloom and presented a most attractive appearance. I examined 20 racemes, but not a flower was punctured. Throughout the season I kept the flowers under close surveillance, but with the same result— none of them were perforated. What was the cause of this result, which was directly opposite to that observed the previous season? For some reason, perhaps the absence of any nests in the vicinity, in 1910, during the entire blooming-period of the scarlet runner, I saw not a single specimen of *Bombus terricola* in my garden, the species of bumblebee so common on the flowers the preceeding season. The honey-bees from the neighboring hives were constantly flying over the garden, but they did not puncture the flowers, doubtless because they were not able to do so. The perforations of the previous season appear thus to have been made wholly by bumblebees. (Fig. 51.)

It has frequently been asserted that honey-bees puncture ripe grapes, but this is undoubtedly an error. The punctures

FIG. 51.   Scarlet Runner.   *Phaseolus multiflorus*

A bumblebee-flower, in which bumblebees sometimes bite holes in order to obtain the nectar more easily

are invariably made by birds or some other agency and are subsequently used by the bees for sucking out the contents of the fruit. Clusters of whole ripe grapes placed in hives among starving bees remained untouched, but if they were pricked with a pin their contents were at once extracted. Honey-bees, however, do bore into soft, succulent tissue for sap. In the common laburnum there is a round, fleshy swelling at the base of the standard, which bees and butterflies pierce for the abundant sap. There are several species of orchis (*O. morio* and *O. maculata*) which are nectarless and which Sprengel called "sham-nectar producers." The inner membrane of the floral tube is a very delicate structure and beneath it there is a copious supply of fluid. Mueller saw a honey-bee pierce this tissue a number of times. Bees also probe with the points of their maxillæ pollen-flowers like the *Anemone*. Moths and butterflies are also able to puncture plant-tissues to some extent. Darwin tells of a moth in Queensland, Australia, which with its wonderful proboscis can bore through the thick rind of an orange. At the Cape of Good Hope both moths and butterflies are said to do much injury to peaches and plums by puncturing the unbroken skins.

Bumblebees are known to bite holes in more than 300 species of flowers and rob them of nectar. Several of these often fail to produce seed. A few of the more common forms robbed by bees, besides those already mentioned, are red clover, locust, *Dicentra*, *Corydalis*, dead-nettle, larkspur, aconite, and vetch. Wasps have also been observed to bite holes in flowers.

Bees are frequently described as roaming about among flowers leading a joyous, care-free existence; but they often meet a terrible fate and are seized by a monster as remorseless as the fabled Scylla of ancient mythology. The *Thomisidæ*, or crab-spiders, have acquired the habit of living among flowers for the

purpose of preying on the insect visitors. They usually lurk
in dense clusters of small flowers, like the inflorescence of the

FIG. 52. Large Insects Captured by Crab-Spiders

1, Butterfly, *Papilio asterias*, captured by crab-spider, *Misumena vatia*, 2. 3, dragon-fly,
Celithemis eponina, killed by crab-spider, *M. vatia*, 4

sumach, meadow-sweet, elderberry, *Viburnum*, cornel, and the
bristly sarsaparilla, although they are also found on large individual flowers like the rose. The commonest species of the

103

family is *Misumena vatia*, a white spider with a crimson stripe on each side of the abdomen, which easily escapes notice until a dead insect is seen lying upon the surface of the inflorescence. Another species (*M. asperata*) has red markings, and sometimes exactly resembles the sorrel (*Rumex Acetosella*).

Misumena does not spin a web, but conceals itself among the flowers and pounces upon its unsuspecting prey while it is collecting pollen or nectar. One morning in July I had the opportunity to observe the capture of a bumblebee gathering pollen on a wild rose. My attention was for a moment diverted, but was again recalled by the loud buzzing of the bee. The spider had leaped upon its back and grasped it with its mandibles just behind the head. At first the bumblebee struggled violently, but so virulent was the poison that its movements speedily ceased entirely. The spider then dragged it over the edge of the flower to the leaves beneath, where it dined at leisure.

The temerity and success with which the *Thomisidæ* attack large butterflies or dragon-flies, or stinging insects, as wasps, bumblebees, and honey-bees, is astonishing. Honey-bees are often captured, and large flies belonging to the genera *Archytas* and *Therioplectes* and rarely the wasp *Vespa germanica*. It is difficult to understand why the spiders are not carried away by such strong-winged insects as the dragon-fly and the large butterfly *Papilio asterias*, which so greatly surpass them in size and strength. (Fig. 52.)

The habit of resorting to flowers to capture anthophilous insects and the protective resemblance of coloration must have been acquired by the crab-spiders in comparatively recent times—that is, since the evolution of flowers and the development of anthophily among insects. The new habit would seem to be the result of observation and experience.

CHAPTER VII

BEES WHICH VISIT ONLY ONE KIND OF FLOWER

ONE warm afternoon on the 20th of July I was collecting insects from a boat on the Medomac River. A thunder-shower was coming up in the northwest. The air was very still and in that peculiar condition which precedes an electric storm. At such times insects are very sluggish and seek shelter against the approaching tempest. The silence was broken only by the rumbling peals of the distant thunder, following the bright flashes of lightning, which illumined the dark thunder-heads of the advancing clouds. It became necessary for me to hasten homeward. To my surprise I noticed on almost every one of the violet-blue spikes of the pickerel-weed (*Pontederia cordata*), a species of water-hyacinth, which in countless numbers fringed the winding stream on both sides, one to several small bees. They had crept within the bilabiate flowers as far as possible, and were evidently intending to await there the passing of the storm. They were so inactive that no net was required, and I could easily knock them off into the cyanide jar. I collected about 40 specimens and could have easily collected hundreds. This phenomenon has never been repeated to my knowledge.

On examination the bee proved to be *Halictoides novæ-angliæ*, or the pickerel-weed bee. Every season when the pickerel-weed is in bloom I find both sexes of this bee on its flowers, and although I have carefully observed the visitors to many other plants in this locality, I have never met with it elsewhere. Apparently in this region it never visits any other flower—it is a monotropic bee. When a species of bee restricts its visits chiefly to one kind of flower it is termed a monotropic bee; or to a few allied kinds of flowers an oligotropic bee; but if it visits many flowers a polytropic bee. These terms were first proposed by Loew, and signify: adapted to one, few, or many flowers.

It is impossible not to feel some curiosity as to why this little bee restricts its visits to the inflorescence of the pickerel-weed. Notice that it flies only at the season of the year when this aquatic plant is in bloom, and that it finds within the perianth both food and shelter. Very likely its nests are built not far away. The flowers of the pickerel-weed strongly attract insects by their great numbers, bright hues, pleasant fragrance and abundant nectar and pollen, and consequently are sought out by many bees, flies, and butterflies. (Fig. 53.) Bumblebees especially delight in these blossoms, which they visit with astonishing rapidity—*Bombus vagans* making about 70 visits per minute. On the middle lobe of the upper lip there are two bright-yellow spots, which tell of the presence and guide to the exact location of the nectar concealed within the tube of the perianth. When the pickerel-weed bee makes its appearance, about the middle of July, there is no other flower in southern Maine which can offer it so many inducements as the pickerel-weed. But let us look further and see if there are any other bees which behave in a similar manner.

In the quiet bays of the river, floating upon the surface of the water, bloom the yellow water-lilies (*Nymphœa advena*). (Fig. 54.)

"Again the wild cow-lily floats
Her golden-freighted, tented boats,
O'ershadowed by the whispering reed,
And purple plumes of pickerel-weed."

106

Fig. 53. Pickerel-Weed. *Pontederia cordata*

In New England a small bee, *Halictoides novæ-angliæ*, never visits any other flower

THE FLOWER AND THE BEE

The flower is securely anchored to the bottom of the stream by a long stem. At first the opening in the bud is no larger than a bee's body, and the chamber within offers a dry and snug shelter amid the waves. It may be truly called a haven of refuge. Directly below the entrance is a broad, many-rayed, crown-shaped stigma, as in the poppy. The petals are thick, wedge-shaped bodies which are orange-yellow on the outer side near the top, where they freely secrete nectar. Under a microscope both large and minute drops can readily be seen. The stamens are indefinite in number; and revelling in the pollen, their bodies completely covered, there is a large and lively company of small flies called *Hilara atra*. Less common are two beetles, *Donacia piscatrix* and *Donacia rufa;* but what chiefly interests us is a small bee, *Halictus nelumbonis*, or the water-lily bee. This bee in this locality is never found on any other flower, but elsewhere it is met with on other species of the water-lily family, or *Nymphœaceœ*. Since, however, it confines its visits to the water-lily family it is an oligotropic bee, and the only species of the great genus *Halictus* that is known to behave in this way.

But in *Andrena* this is a common phenomenon; for instance, in Washington County, Wis., according to Graenicher, 24 of the 47 indigenous species of *Andrena* are oligotropic. This is the largest genus of North American bees. They are sometimes called ground-bees, since they build branched tunnels 8 or 10 inches deep in the soil of sandy pastures and hillsides. A part of the species are vernal or fly in springtime, while a part are autumnal and fly only in autumn. They provision their cells with balls of "bee-bread," about the size of a garden-pea, composed of pollen moistened with nectar. An egg is laid on the top of the mass of bee-bread, and the cell is then closed.

FIG. 54. Yellow Water-Lily. *Nymphœa advena*
Largely pollinated by small flies, *Hilara atra*

THE FLOWER AND THE BEE

The bright-yellow staminate aments of the pussy-willow (*Salix discolor*) (Fig. 55) are great favorites of vernal species of *Andrena*, whence Smith calls them "harbingers of spring." The pussy-willows bloom in northern New England during the latter part of April, and their bright-yellow aments are very pleasing objects in the cold, gray landscape. They are very attractive to a varied company of insects, as honey-bees, bumblebees, flies, butterflies, and beetles. It is a busy scene and one which the naturalist can never tire of watching; but it is not one of unmixed happiness, for little tragedies take place before our eyes. Among those which come to sip the nectar are little dance-flies (*Empididæ*), and not infrequently they are seized and carried away bodily by black robber-ants which roam everywhere. Honey-bees and many species of *Andrena* come in great numbers to procure pollen for brood-rearing. A part of the andrenid bees gather only a portion of the pollen they require from the willows and the balance from the maples, plums, cornels, and *Viburnums;* but there are four species (*A. illinoiensis*, *A. mariæ*, *A. erythrogaster*, and *A. moesta*), which get their whole supply from this genus of plants. Of the autumnal flying species of *Andrena* there are five (*A. canadensis*, *A. nubecula*, *A. solidaginis*, *A. hirticincta*, and *A. asteris*), which I have collected only on the flowers of the *Compositæ*, or aster family; and four of these bees confine their visits very largely to the goldenrods. In both *Salix* and *Solidago* the inflorescence offers an ample supply of nectar and pollen, and there is little temptation for andrenid bees to go elsewhere, when their time of flight coincides with the period of blooming of these two genera.

But in other localities *Andrena erigeniæ* is reported to be a monotropic visitor of the spring-beauty (*Claytonia virginica*), and *Andrena violæ* of the violet (*Viola cucullata*), *Andrena*

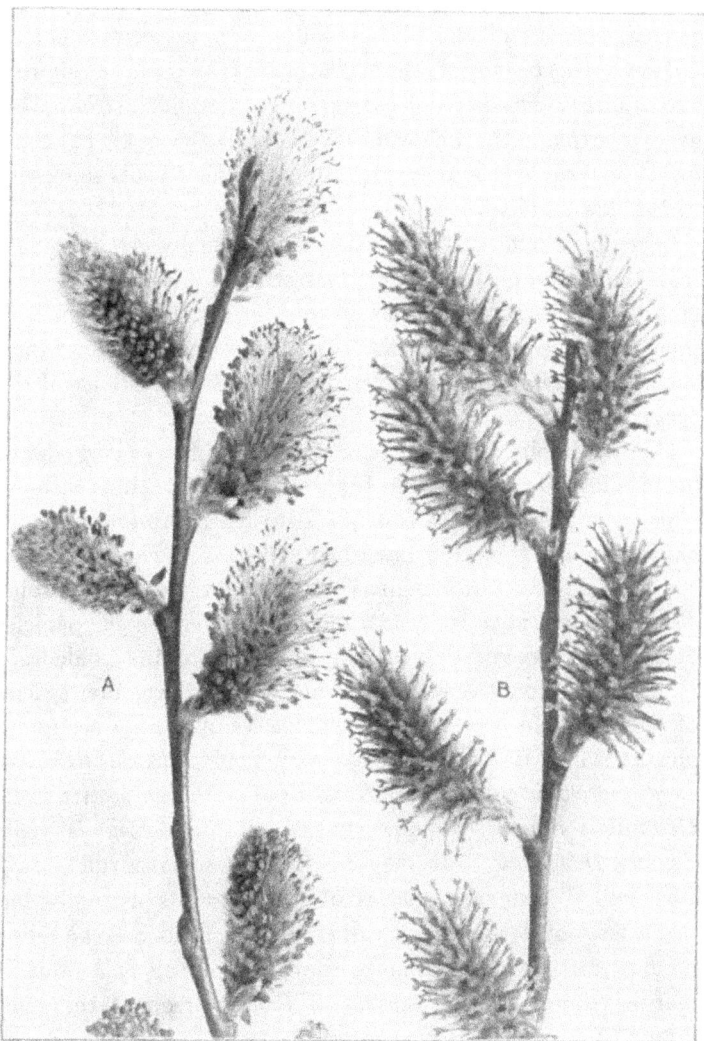

FIG. 55. Pussy-Willow. *Salix discolor*

In New England four species of ground-bees (*Andrena*) never visit any flowers except those of the willows.   *A*, staminate catkins; *B*, pistillate catkins

geranii maculati of the wild geranium (*Geranium maculatum*), *Andrena fragariana* of the strawberry (*Fragaria virginica*), and *Andrena parnassiæ* of *Parnassia caroliniana*. It is not so easy to explain the behavior of these latter bees. It seems very remarkable that they should restrict their visits so closely to the flowers mentioned.

Macropis ciliata, or the loosestrife-bee, usually gets its pollen from the flowers of the common loosestrife (*Lysimachia vulgaris*) (Fig. 97); but it visits other flowers for nectar with which to moisten the pollen, since the loosestrife is nectarless. Many species of *Panurginus* are taken only on the inflorescence of the *Compositæ*.

But the habit of visiting only one kind of flower is, perhaps, better illustrated by *Perdita* than by any other genus of bees. This large genus of bees is confined to North America and includes not far from 150 described species and varieties, most common in the arid regions of New Mexico. In Maine *Perdita octomaculata* is found almost exclusively on the panicles of *Solidago juncea*, the earliest blooming of the goldenrods (Fig. 56), and only very rarely is met with on any other species of *Solidago*. In New Mexico two species of *Perdita* are found on the willows, *Perdita zebrata* visits only *Cleome serrulata*, *Perdita crotonis* visits *Croton texensis*, *Perdita albipennis* visits *Helianthus annuus* (sunflower), and *Perdita senecionis* visits *Senecio Douglasii*. "It may be laid down as a rule," says Cockerell, "that each species of *Perdita* visits normally but one species of flower, but occasionally specimens may be found on flowers to which normally they do not belong." But in many instances several species of *Perdita* frequent the same flower.

Many species of *Colletes*, *Epeolus*, and *Melissodes* visit almost exclusively the flowers of the *Compositæ*, as the thistles, golden-

Fig. 56. Early Goldenrod. *Solidago juncea*

rods, and asters. *Xenoglossa pruinosa* confines itself to *Cucurbita Pepo*, or the common field-pumpkin; while *Megachile campanulæ*, one of the leaf-cutting bees, is a monotropic visitor of the bellflower *Campanula americana*. Many other instances are recorded, and many more will no doubt be discovered when our bee fauna is better known.

This is certainly a very singular habit on the part of bees, and one which few would be likely to foresee. On the contrary, it is generally supposed that bees fly about sipping sweets indiscriminately, as they are so commonly represented by the poets.

> "He woos the Poppy and weds the Peach,
>   Inveigles Daffodilly,
> And then like a tramp abandons each
>   For the gorgeous Canada Lily."

It is really getting unsafe for poets to write about Nature in their old haphazard way, trusting chiefly to their imagination as a guide. Fancy can supply nothing half so wonderful as the true facts about flowers and insects. Let us consider what theories naturalists have advanced to explain this curious habit.

In Kerner's day only a few oligotropic bees were known, and he believed that they gave the preference to certain flowers because they found their odor so highly attractive. But it is incredible that so many bees should be dominated in their flight to such an extent by various floral odors, and besides they not infrequently visit several flowers which differ in scent. No doubt, though, bees have their preferences in odors and nectars, and probably they prefer pollen that has a roughened or spined surface to that which is smooth.

Another explanation claims that the bee fauna is as large as the flora will support and that female oligotropic bees have adopted this method of visiting flowers to avoid competition in gathering pollen for brood-rearing. But this is not the fact and it can be shown that only a part of the available flower-food is gathered by bees. The commonness of an insect species does not depend alone on the quantity of food obtainable, *e. g.*, occasionally the forest-caterpillar (*Heterocampa guttivitta*), which feeds on the leaves of deciduous trees, appears in countless numbers, defoliating acres of the woodlands and apparently threatening the entire destruction of the hardwood forest; but it speedily disappears again and becomes so rare that its presence is unnoticed. The size of the bee fauna is likewise limited by other factors than the food supply, the most important being insect parasites which destroy annually vast numbers of bee larvæ.

It is desirable to consider a few specific instances where there is unquestionable evidence of a surplus of flower-food. In Riverside County, Cal., the orange-bloom secretes nectar so freely that it drips upon the clothing of the pruners, and at the end of a day's cultivating in the groves it is necessary to wash the horses and harnesses. Large quantities are lost each year for want of bees to collect it. Hundreds of acres of the sandy, coastal plain of Georgia are covered with the bushes of the common gall-berry (*Ilex glabra*). It remains in bloom for about a month. The secretion of nectar is constant and but little affected by the weather; but this sea of flowers is not frequently visited by insects. Immense quantities of fine honey are lost annually because there are no bees to gather it; furthermore, it is not easy to overstock a gall-berry region with the domestic bee, and in one instance 362 colonies did nearly as well as 100 previously. The production of honey in Iowa is placed at ten to twelve million pounds annually; but according to a conservative estimate by Iowa apiarists of great experience, it is possible to produce in that State in a single year 60,000,000 pounds. The average moisture content of honey is 17.59 per cent, while that of nectar is not far from 75 per cent, so that the weight of the nectar would exceed that of the honey fourfold. This estimate, of course, does not take into consideration the nectar consumed by anthophilous insects other than the honey-bee. If a region were already stocked to its fullest capacity with bees, it is clear that it would be impossible to establish large apiaries containing millions of bees, storing twenty or more tons of honey, consuming, perhaps, twice as much more, and requiring enormous quantities of pollen for brood-rearing. It will be remembered that the honey-bee does not usually fly more than two miles from the apiary.*

It would be easy to multiply examples in the case of buckwheat, basswood, tupelo, raspberry (Fig. 57), heart's-ease, and goldenrod, and the extra-floral nectaries of cotton and *Cassia Chamæchrista* in the Southern States. Certain plants, as *Bidens aristosa* in the lowlands of the Mississippi, fairly carpet large areas with their myriads of flowers. The "Big Sawgrass" is a tract of land in Florida covering a thousand acres. It is a wilderness of weeds, a dense jungle of grass and flowers with vast stretches of nectariferous plants, like boneset and wild sunflower, yielding honey enough to keep a thousand colonies busy for months; but as yet there are only fifty colonies in one

* The question might be raised at this point whether there are not too many bee-keepers already, or whether the ten million colonies are not using all the honey or nectar there is in flowers. The facts are, more nectar goes to waste than is gathered. It has been estimated that from 50 to 80 per cent of it is lost simply because there are no bees in the vicinity to gather it. It is at least conservative, says Doctor Phillips, apicultural expert of the Department of Agriculture, in his book, *Beekeeping*, to say that ten times as much honey could be produced in localities where there are now no bees or an insufficient number, as is now produced. In other words, the resources of this country could furnish $200,000,000 worth of honey instead of $20,000,000, as at present.—E. R. Root, in *A B C and X Y Z of Bee Culture*, p. 3.

FIG. 57. Wild Raspberry. *Rubus idæus* var. *aculeatissimus*
Yields nectar very freely, and is annually the source of many tons of a delicious white honey

corner. What a wealth of sweetness going to waste! Fruit-growers have learned from experience that the wild bees are wholly insufficient to gather the pollen and nectar of extensive plantations of fruit-trees, berry-bushes, and cranberries; and effective pollination is secured only by the establishment of apiaries of the domestic bee. An immense quantity of pollen, which can be used by bees in emergencies, is produced by ane-mophilous plants, as the *Amentaceæ*, elms, grasses, sedges, rushes, and a variety of homely weeds. Occasionally honey-bees by thousands do resort to anemophilous flowers for pollen; and much less frequently, because their necessities are less, the solitary bees. Many plants have probably remained wind-pollinated, while others formerly entomophilous have wholly, or in part, reverted to self-fertilization or anemophily in the absence of sufficient pollinators.

If severe competition did exist among the solitary bees for flower-food, the oligotropic habit would not be desirable. It is not an advantage for a bee to restrict its visits to one kind of flower unless it is always certain to obtain the food supply it requires; otherwise it is clearly at a disadvantage as compared with the polytropic species. If severe competition is introduced by artificial means, as by overstocking a locality, then the oligotropic bees will either tend to disappear or become polytropic. The small number of oligotropic bees reported from central Europe is noteworthy. If, however, a very com-mon flower yields a surplus of food then a bee with a period of flight nearly coinciding with the period of inflorescence would save time and labor by restricting its visits to this species; and since bees instinctively learn from observation it would naturally be expected that the oligotropic habit would be formed. According to the theory proposed by the writer certain bees have become oligotropic because of the direct advantage gained, combined with a short term of flight, or a

BEES WHICH VISIT ONLY ONE KIND OF FLOWER

flight synchronous, or nearly so, with the period of inflorescence of the plant to which they restricted their visits. This theory offers an explanation of the rise of oligotropism by the observation of existing conditions. There may be, and often are, accessory factors, as small size, time of flight, length of flight, weak flight, vicinity of nests, and the number of bees.

The relation of the domestic bee to various flowers affords an ever-present illustration of the way in which the oligotropic habit might arise in the case of a bee with a short term of flight. While the basswood and white clover are in bloom the honeybee visits these flowers almost exclusively. Again, in the fall in Maine it confines its attention solely to the goldenrods. In California at times it collects nectar exclusively from the sages; in Michigan from the willow-herb, and in other regions from other plants. If from any one of these plants it also obtained its supply of pollen and was on the wing only while it was in bloom, it would be a monotropic bee in the strict sense of the word. There is here no question of competition; the bees come to procure the great abundance of nectar, and pollen is gathered at the same time as a matter of convenience. But where a bee flies from early spring to late fall and requires a large amount of stores, it is evident that it can never become oligotropic.

There can be no competition where there is an overabundance of supplies. No other early blooming flowers yield so much pollen and nectar as the willows. No other genus of honeyplants in early spring is so valuable to the apiarist as *Salix*. Honey-bees gather large quantities of pollen, and in some localities are reported as storing from 8 to 15 pounds of honey per hive from this source alone. Four species of *Andrena*, which are on the wing for about a month, visit the willows exclusively, because during their comparatively short term of flight they can readily obtain all the pollen and nectar they re-

119

quire, and there is no occasion for them to go elsewhere. There are also on the wing at the same time 6 species which are polytropic, but they all obtain a part of their food supplies from the willows, so that the oligotropic species would not escape competition with them if there was a scarcity of pollen and nectar. Their average time of flight is about sixty-three days, or 43 per cent longer than that of the oligotropic species, which renders it necessary for them to obtain a part of their pollen from other flowers than those of *Salix*.

In Milwaukee County, Wis., according to Graenicher, there are 11 autumnal species of *Andrena;* and all of them, with one exception, are oligotropic visitors of the *Compositæ*. The single exception (*A. parnassiæ*) is found only near Whitefish Bay, Lake Michigan, where *Parnassia caroliniana* produces a great abundance of flowers. Evidently this bee gets its pollen from these flowers because they are very common in the one locality where it is known. All the other species, 10 in number, are oligotropic to the *Compositæ*. Many genera of this family are exceedingly common, as the goldenrods, asters, sunflowers, and thoroughworts, and yield immense quantities of nectar and pollen. There are very strong inducements for these bees to visit these flowers, and comparatively little for them to go elsewhere. In New England 4 species of *Andrena* restrict their visits to the goldenrods, from which the honey-bees gather annually many tons of honey and a great amount of pollen. Neither the visits of the domestic bee nor of the andrenid bees are the result of competition, but solely of the advantages gained.

The majority of oligotropic bees flying in summer and autumn, whether they be species of *Colletes, Andrena, Perdita, Panurginus,* or *Melissodes,* visit exclusively the *Compositæ*. The large and crowded inflorescence consisting of many small flowers which can be quickly and easily visited, the great abundance of pollen and nectar, and the commonness and wide distribution of many species are the factors which attract these bees. No other family of plants blooming at the season offers equal advantages. The different genera of the *Compositæ* vary greatly in the length of the floral tubes, while in the genera of bees the length of the tongue also varies greatly. Thus it is the tube length of the flower which is the factor limiting the visits of many species of bees to certain composite genera. Small bees with short tongues do not resort to the same flowers as larger bees with longer tongues. (Fig. 58.)

Practically all of the species of *Perdita* are oligotropic. They are small bees with a short flight. A part of the species are vernal; but the majority fly in late summer and autumn and many visit the *Compositæ*. The flowers visited by them occur in immense profusion and include the best honey-plants of this country, as *Salix*, *Solidago*, *Cleome*, *Prosopis*, *Helianthus*, *Verbesina*, *Bidens aristosa*, and *Monarda punctata*. It is noteworthy that we find these flowers also visited by oligotropic bees belonging to other genera. This behavior on the part of so many species of bees is very similar to that of higher forms of life when they gather at some feeding-ground where there is a superabundance of food.

Since the nest-building bees are compelled to collect pollen for brood-rearing they are naturally more constant in their visits to flowers than the parasitic bees, which do not gather pollen and require nectar only for themselves. Nevertheless a number of the parasitic bees with a short term of flight visit wholly or largely the *Compositæ* and may be regarded as oligotropic species. This is of great interest since it is not claimed that *they* have acquired this habit as the result of competition.

We may sum up the matter as follows. All bees including the honey-bee show a strong tendency in collecting both nectar and pollen to be constant to one species of flower. This is manifestly for the advantage of both insects and flowers. In the case of a number of bees flying for only a small part of the season this habit has become so specialized that they visit only one or a few allied species of flowers, which offer an abundance of pollen and nectar. As the honey-bee for a time restricts its visits to the white clover, so in like manner a monotropic bee visits but a single kind of flower. But in the former case the bee flies throughout the whole season; but in the latter, when the flower fades, the bee's period of flight is over.

The idiosyncrasies of bees in visiting flowers present many remarkable peculiarities, and undoubtedly offer an attractive field for observation. There are certain bees which, though they are not oligotropic, obtain the larger part of their supplies from comparatively few flowers, as the plums, thorn-bushes, cornels, and viburnums. In this locality one of the leaf-cutting bees (*Megachile melanophœa*) shows a decided preference for the purple vetch (*Vicia Cracca*), and if I desired a specimen I should look for it on the blossoms of this plant. Since the male bees do not gather pollen they may not visit the same flowers as the females, though the attraction of the female may largely influence their course, in which respect they exhibit quite human sentiments. It would, of course, be in vain to look for the males of *Bombus* and *Halictus* on the flowers of spring, since they do not appear until midsummer. In the case of diœcious plants, or plants in which the sexes are on different individuals, the bees visiting the staminate flowers are more numerous and are sometimes widely different from those visiting the pistillate. The common sumac is a good example. Indeed, the bees visiting a flower in its early stages may differ from those visiting it in its later stages. Again the visitors to a flower may differ, both in number and kind, in different seasons.

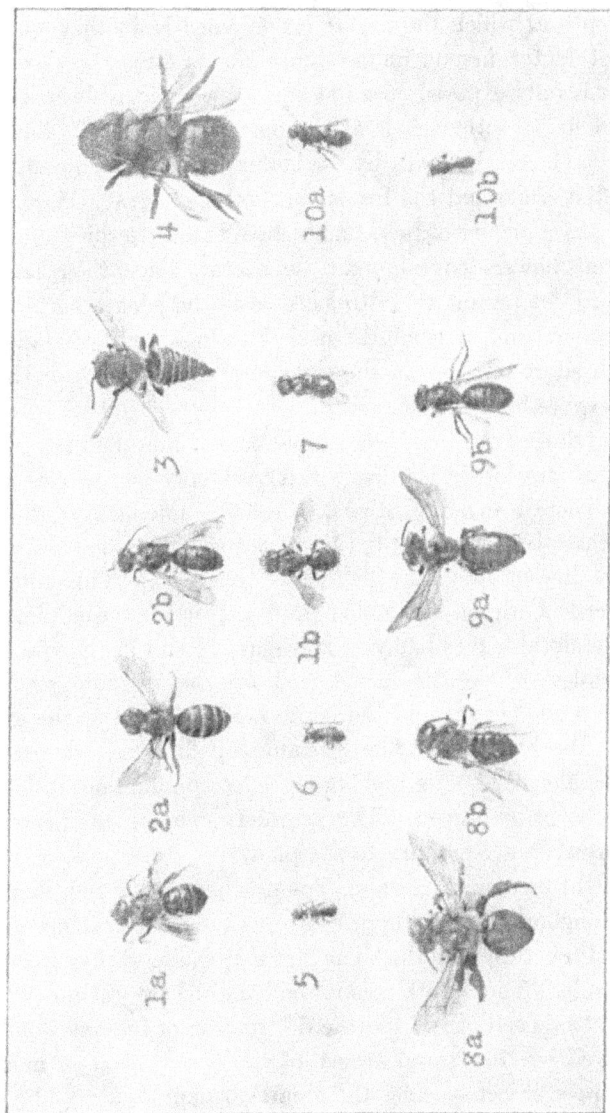

FIG. 58.   Common Solitary Bees of the Eastern States

1, *Halictus craterus*, *A*, female, *B*, male; 2, *H. lerouxii*, *A*, female, *B*, male; 3, *Coeliozys rufitarsis*, female; 4, *Osmia megacephala*, female; 5, *Halictus hortensis*, female; 6, *H. virecki*, female; 7, *Osmia pumila*, female; 8, *Andrena vicina*, *A*, female, *B*, male; 10, *Prosopis modesta*, *A*, female, *B*, male

THE FLOWER AND THE BEE

The depth at which the nectar is concealed is another most important factor in controlling the visits of bees. In some flowers it is fully exposed on a flat surface where it is accessible to all insects; in others it is at the bottom of a slender tube, where it can be reached only by the larger moths. The familiar fable of the crane and the fox is constantly illustrated among flowers. As a matter of fact, bumblebees and butterflies avoid rotate, flat flowers containing little nectar, since their long tongues do not permit them to suck easily on such a surface. On the other hand, it would be useless to look for the smaller bees with short tongues on the larkspurs and clovers, for the nectar is quite beyond their reach.

As we take leave of the oligotropic bees it may be inquired if there are any other insects, which visit only one species of flower. There are many others, especially among butterflies and moths. The flag-beetle (*Mononychus vulpeculus*) passes its entire life on the blue flag (*Iris versicolor*). This small weevil feeds both on the pollen and nectar, and sometimes gnaws the floral leaves badly. The eggs are laid in the young seed-capsules, where the larvæ feed on the ripening seeds. Both the adult beetles and the larvæ are supported at the expense of the blue flag. The legitimate pollinators are bees, and while the flag-beetle may rarely effect pollination, it does far more harm than good. This symbiotic relation is a benefit to the insect, but an injury to the plant.

The night-blooming yucca, or Spanish bayonet, which flourishes throughout the Southern States, is pollinated exclusively by a small nocturnal moth. The larvæ of the moth live in the seed-capsule. Thus both plant and moth are reciprocally dependent on each other, and the destruction of the one would be followed by the disappearance of the other. But in most instances the insect receives the greater benefit.

CHAPTER VIII

BUTTERFLY–FLOWERS

FLOWERS are the playground of butterflies, where all day long in warm, sunny weather they flit about "like tickled flowers with flowers." Pirouetting the hours away in airy dances and free from the care of providing for their offspring, butterflies are permitted by Nature to enjoy a greater amount of pleasure than is granted to any other of her insect children. Their beautiful colors have made them the favorites of collectors, and led Jean Paul to call them "the flowers of the air." When they migrate in vast numbers, as they occasionally do, they fill the air with clouds of color comparable to the unbroken sheets of bloom displayed by the mountain-laurel and the flame-colored azalea. The beauty and brilliancy of the bird-winged butterflies (*Ornithoptera*) in the oriental tropics, says Wallace, are indescribable; and the capture of a new species filled him with such intense joy that on taking it out of his net and opening its glorious wings, his heart began to beat violently, and he felt much more like fainting than he had done when in apprehension of immediate death.

In the forests of the Amazon, says Bates, brilliant-hued butterflies occur in so great numbers and in such endless diversity that they compensate for the scarcity of flowers and are a feature in the physiognomy of the landscape. On the moist sand-beaches of the river vast numbers of sulphur-yellow and orange butterflies congregate in densely packed masses two or three yards in circumference and resemble beds of yellow crocuses; while flitting about among the trees a butterfly with wings as transparent as glass, except for a spot of violet and rose, looks like "the wandering petal of a flower." In the variegated patterns of their wings, he declares, Nature writes as upon a tablet the story of the modification of species. "Therefore, the study of butterflies—creatures selected as types of airiness and frivolity—instead of being despised, will some day be valued as one of the most important branches of biological science."

Bates's words were prophetic, and it is to-day generally recognized that the brilliant markings on the wings of butterflies offer to students of evolution and heredity a most promising field for investigation. Eimer, who based his theory of orthogenesis chiefly on these patterns and colorings, says: "Like the leaves of an open book the written characters on the wings of our butterflies show their past and present history."

According to the catalogue published by the United States National Museum, in 1902, there are in North America 6,622 species of *Lepidoptera*, of which 652 are butterflies and 5,970 are moths. While in other orders of insects a part of the species live upon other substances than nectar and pollen, all of the *Lepidoptera* are adapted to a floral diet. But since they do not collect pollen but feed on nectar alone, they are far less important than bees as pollinators and much less constant in their visits. Doubtless all the butterflies at times visit flowers, but only about 107 have actually been collected on flowers in North America and 111 in Europe. Many adult moths do not take any food at all, so that with the exception of the hawk-moths and a few other families, notwithstanding their great numbers, they are not frequently observed sucking nectar. In North America, including 25 species of hawk-moths, or *Sphingidæ*, only 82 have been listed as flower-visitors; and in Europe 186, of which 36 are hawk-moths. The tubular proboscis or

Fig. 59. Sweet-William. *Dianthus barbatus*

A butterfly-flower

tongue, which is carried coiled beneath the head, varies in length from $\frac{1}{20}$ of an inch to more than 10 inches; and is formed by the extension of the blades of the maxillæ, which are held together by minute hooks so that it is practically air-tight.

Among butterfly-flowers none are more widely known than the pinks. They exhibit a wonderful variety of red shades, varying from white, through rose, pink, and deep red to scarlet and crimson. The petals may be marbled or dotted with white, with a white centre, surrounded by a purple ring, as in *Dianthus deltoides*. The corolla is often notched or fringed and surmounted by a corona of scales. The perfume is aromatic, and the nectar is deeply concealed. The red-flowered pinks are adapted to pollination by butterflies by which they are chiefly visited.

The variegated flowers of the sweet-william, or bunch-pink (*Dianthus barbatus*), familiar in every flower-garden, display the most vivid shades of crimson and scarlet and, as the name indicates, exhale a pleasant fragrance. (Fig. 59.) They are adapted to pollination by butterflies and day-flying moths. The nectar lies at the bottom of a long calyx-tube beyond the reach of honey-bees, which I have seen vainly thrusting their tongues down the centre of the flowers, probing between the petals, and even looking under the corolla.

The carmine flowers of the stemless catchfly (*Silene acaulis*), which grows in the higher Alps, are very frequently visited by butterflies, upon which they are dependent for pollination. Two species of *Lychnis* have beautiful bright-red flowers, which are very attractive to butterflies. Twenty-eight different species of butterflies have been taken on the handsome, red flowers of the soapwort (*Saponaria ocymoides*); the pinks (*Dianthus*) also have the nectar so deeply concealed that it can be reached only by *Lepidoptera*, a part of the elegant red flowers

Fig. 60. Orange-Red Lily. *Lilium philadelphicum*
A butterfly-flower

THE FLOWER AND THE BEE

being adapted to butterflies, and a part to diurnal hawk-moths. "As the honey gets more deeply concealed and access more directly limited to butterflies, we find," says Hermann Mueller, "*pari passu* among the *Caryophyllaceæ* (pink family) increasing development of sweet scents, bright-red colors, fine markings round the entrance of the flower, and indentations at the circumference. All these characters, which are so attractive to us, seem to have been produced by the similar tastes of butterflies." This conclusion is much strengthened by the fact that nocturnal flowers are usually white and without variegation.

The wild orange-red lily (*Lilium philadelphicum*), which grows in dry, upland pastures, is pollinated by butterflies (Fig. 60), while the wild yellow lily (*Lilium canadense*), which blooms along the marshy river-banks, is pollinated by bees. The bee-lily is an inverted, bell-shaped flower with broad overlapping floral leaves, which shed the rain perfectly. Bees alight on the stigma and crawl up the style to the nectar at the bottom of the flower. (Fig. 61.)

But the butterfly-lily stands erect, and the floral leaves are contracted at base into narrow claws, leaving wide interspaces through which the rain easily escapes. If the perianth formed a cup, like that of the bee-lily, it would fill with water; and, if it were inverted, it could not be conveniently visited by butterflies. I never fail to watch with pleasure the manœuvres of butterflies to obtain the nectar of this wild lily. The narrow claw of each perianth segment has its edges turned inward to form a groove, which guides the proboscis of the butterfly to the nectar-gland at its base. The only visitor I have observed is the common, yellowish-red butterfly *Argynnis aphrodite*. (Fig. 62.) Alighting on the broad limb of the flower, it runs its tongue down one of the grooves to the nectar, while at the same time its wings come in contact with the anthers and stigma. The anthers are covered all around with pollen and are versatile, that is, they are attached by a middle point and when touched oscillate easily up and down like the walking-beam of an engine. In the bee-lily the anthers are fixed in

FIG. 61. Canada Lily. *Lilium canadense*
A bee-lily

one position. A third species of lily (*L. Martagon*) is adapted to hawk-moths.

Lepidopterid, or butterfly and moth flowers, are not numerous, and in the whole Alpine flora Mueller found but 33. Besides the pinks and lilies already mentioned, several red-flowered species of *Phlox* (Fig. 63), a crimson heath (*Erica carnea*), and 5 or 6 red-flowered primroses are pollinated by butterflies. Many lepidopterid flowers occur among the orchids, and in the genus *Habenaria* the beautiful, purple-fringed orchis is a butterfly-flower (Fig. 64), while the greenish or white species are pollinated by crepuscular or nocturnal moths. In some instances I have found the grayish hairs of moths adhering to glutinous surfaces.

Butterflies do not confine their visits to butterfly-flowers alone, but may visit any flower. They experience, however, more or less difficulty in sucking nectar on flat surfaces and consequently prefer tubular flowers with concealed nectar—the longer the tongue the more marked this preference becomes. They also occasionally fly to pollen-flowers and search them for sweet secretions. But no flowers are so frequently visited by butterflies as social flowers of the type of the *Compositæ*, to which 40 to 60 per cent of their visits are made, or 3 to 6 times as many visits as are made to butterfly or bee flowers. Every one has observed a cloud of butterflies hovering over a clump of purple thistle-heads, or the yellow flowers of the elecampane (*Inula Helenium*), or the dull-white clusters of the thorough-wort. The male butterflies, which are often pleasantly scented, pursue the females from flower to flower without any regular order.

Butterflies often rob flowers of their nectar without rendering any service in return. Both honey-bees and butterflies steal thousands of pounds of alfalfa nectar annually through a crevice in the side of the flower. Many other leguminous flowers are robbed in the same way, but in many species the petals close up so firmly that access to the nectar can be gained only in the normal way. While butterflies cannot pollinate the flowers of the blue flag (*Iris versicolor*), they often stand on the upper or under side and, inserting their tongues side-

132

FIG. 62. Orange-Red Butterfly. *Argynnis aphrodite*
The butterfly which pollinates the orange-red lily

ways between the sepals and petaloid styles, suck the nectar. Indeed, it may easily happen in the case of some irregular flowers, as the larkspur, that butterflies may visit them normally and obtain the nectar, and yet not touch either the anthers or stigmas with their slender tongue.

Butterfly-flowers, as has been previously pointed out, are commonly red-colored. Among the Alpine butterfly-flowers which have red corollas are *Orchis globosa, Lilium bulbiferum,* the heath *Erica carnea;* the pinks, *Dianthus superbus, D. sylvestris, D. atroruber; Daphne striata,* and *Primula acaule,* and several other primroses. Other red butterfly-flowers are species of *Silene, Lychnis, Asclepias,* and *Monarda.* "It is certainly not purely accidental," says Mueller, "that most of the butterflies of the Alps, the commonest floral guests in that region, are vivid-red in color, and that bright-red flowers are visited with marked preference by such butterflies." Mueller further observed that orange-hued composite flowers, such as hawk's-beard (*Crepis aurea*) and orange hawkweed (*Hieracium aurantiacum*) are a veritable playground in sunny weather for butterflies of fiery-red color. Two copper-colored butterflies were also observed to fly to the bright-red fruits of the sorrel. This remarkable correlation certainly deserves careful consideration by students of the color-sense of insects.

There would seem to be no *a priori* reason why red butterflies may not be strongly influenced by red flowers. The ornamental coloring of their wings is largely the result of sexual selection; and, since the different sexes readily recognize each other, it is not improbable that in seeking nectar they are specially attracted by flowers of the same color as themselves. This view is strengthened by the fact that blue butterflies may show a preference for blue flowers, *e. g.,* blue species of *Lycœna,* have been seen to favor the blue blossoms of *Phyteuma.*

FIG. 63. Red Phlox. *Phlox paniculata*
A butterfly-flower

THE FLOWER AND THE BEE

Red waves of light, as is well known, excite attention and are seen where other hues are passed by unnoticed; they are the longest waves of the solar spectrum and, like long oceanic waves, possess a great amount of energy. Are not red flags constantly used for signals, and do not soldiers to-day avoid wearing scarlet uniforms? In moderation red is a warm, stimulating color, and is frequently used in wearing apparel, in pictures, and in the decoration of walls and ornaments; but in excess it produces irritation and anger. It enrages the turkey gobbler of the farmyard and excites the Texas steer to madness. Physicians tell us that people living continually in bright-red rooms are apt to be irritable and quarrelsome, but that when the walls are painted a quieter hue, as pale blue or drab, these nervous symptoms speedily disappear.

Edible berries are more often red than any other color. Bird-flowers are almost invariably fire-red or scarlet. In tropical America, where there are more than five hundred species of humming-birds, there are scores of scarlet bird-flowers, as scarlet sages, fuchsias, and abutilons; while in Europe, as Kerner has pointed out, where neither the humming-birds of America nor the sun-birds of Africa nor the honeysuckers of Australia are found, scarlet blossoms are noticeably absent. It is difficult not to believe that anthophilous birds have learned to associate bright-red colors with the presence of an ample food-supply of nectar and small insects.

It is, on the other hand, noteworthy that in the families and genera which contain red butterfly-flowers blue is very rare or wholly absent. There are no blue flowers in the pink family, and in the orchis family out of 6,000 species there is only 1 blue form, *Vanda cœrulea*, from India. Neither are all butterfly-flowers red, for in the genus *Globularia* there are 3 bright-blue species which are adapted to butterflies, "the only in-

FIG. 64. Purple-Fringed Orchis. *Habenaria psychodes*
A butterfly-flower

stance in the German and Swiss flora of a blue color being produced by the selective agency of the *Lepidoptera*." Butterflies, moreover, do not confine themselves chiefly to red flowers, but visit a great variety of colors, and of 1,432 visits made by 100 species, 45 per cent were made to yellow and white flowers, and 55 per cent to red and blue flowers. The dingy-white flowers of the thoroughwort are often alive with butterflies. Finally butterflies visit the level-topped inflorescence of the *Compositæ*, which offers good landing-places and abundant nectar in slender tubes, more frequently than any other flowers, which would show that they were influenced by form more than by color. It seems, therefore, an open question whether the red coloration of butterfly-flowers is not largely an incidental result, rather than due to the selective agency of butterflies.

CHAPTER IX

NOCTURNAL OR HAWK–MOTH FLOWERS

FLOWERS which bloom in darkness seem weird and unnatural. Most conspicuous blossoms are creatures of sunshine and warmth, and seek to allure diurnal insects, while many of them close at the approach of night. But nocturnal flowers are adapted to pollination by moths, chiefly hawk-moths. How this reciprocal relation became established it would be hard to tell; but their forms, time of opening, and colors easily distinguish them from the day-bloomers.

Consider, for instance, the thorn-apples (*Datura*), which have long, slender corolla tubes some six inches in length. (Fig. 65.) They are "children of the dewy moonlight," and fill the evening air with their sweet fragrance. Their large, pale, salver-shaped blossoms "serenely drooping awaken visions of silent awe," and it is at once apparent that these stately flowers do not invite the visits of bees. Some fifty years ago Felicia Hemans was a popular poet in New England, and while she probably knew nothing of the mysteries of flower-pollination, in her lines to *Datura arborea* she instinctively recognizes the fact that bees are not found in this domain of shadows:

> "Majestic plant! such dreams as lie
> Nursed, where the bee sucks in the cowslip's bell,
> Are not thy train:—those flowers of vase-like swell,
> . . . worthy, carved by plastic hand,
> Above some kingly poet's tomb to shine
> In spotless marble."

In their relations to flowers moths may be divided into two groups, the highly specialized hawk-moths (*Sphingidæ*) and the other moth families. Many moths fly only on the rainiest and darkest nights. We should like to know more of the devious ways of these nocturnal wanderers amid the down-pouring rain. They seem a bit uncanny. Among the smaller moths most frequently observed on flowers are the measuring-moths (*Geometridæ*), the leaf-rollers (*Tortricidæ*), the owlet moths or noctuids (*Noctuidæ*), and the teneids or the little moths of the family *Teneidæ*, the larvæ of which mine in leaves. Few of them are common floral visitors, or of much significance in pollination. Several of the hawk-moths, as the clear-winged moths, fly regularly in the daytime.

The yuccas, or Spanish bayonets, liliaceous plants which are widely distributed in this country and Mexico, are entirely dependent for pollination on little teneid moths of the genus *Pronuba*. If the phenomena attending the transfer of the pollen had not been investigated by Riley and Trelease in every detail they would seem as incredible as a tale of Munchausen. The large, pendulous flowers are creamy-white tinged with green or rose, and are borne in magnificent clusters or panicles, which are well worthy of the admiration they excite. Subtropical species of this genus become arboreal and reach an altitude of 30 feet. In California *Yucca Whipplei* sends up a flower-stalk 12 feet high, which for nearly half its length bears an imposing cluster of flowers. (Fig. 66.)

Since the large, bell-shaped flowers hang downward and the stigmas stand in advance of the anthers, self-pollination is impossible, for the pollen is too glutinous to be carried by the wind, and if accidentally dislodged, it falls directly to the ground. The continued existence of the yuccas, therefore, depends chiefly on the little moths of *Pronuba*. The female

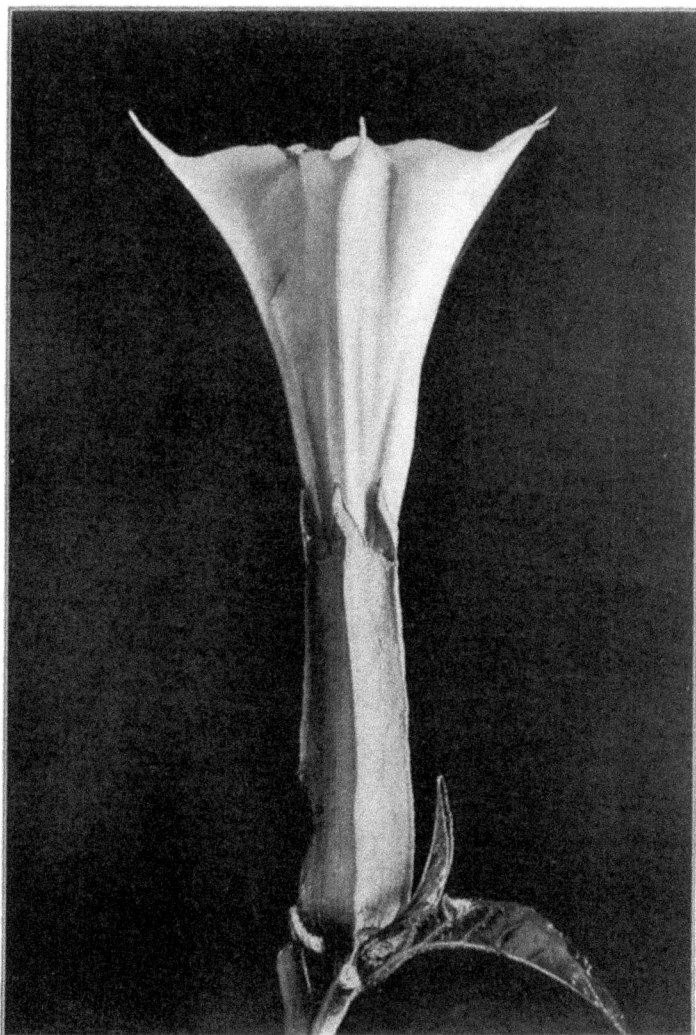

Fig. 65. Thorn-Apple. *Datura Tatula*

A hawk-moth flower

FIG. 66. *Yucca Whipplei* of California

Twelve feet tall. The magnificent cluster of flowers is about six feet long. (After Riley)

NOCTURNAL OR HAWK-MOTH FLOWERS

Pronuba is unique among all the thousands of moths and butterflies in the world in that she has maxillary tentacles for collecting pollen, and a horny ovipositor for piercing succulent tissue. In the collection of pollen she resembles the bees, and in the manner of laying her eggs the ichneumon-flies—both hymenopterous insects.

The most widely cultivated and best-known species of yucca is Adam's needle, or Spanish bayonet (*Y. filamentosa*), which is pollinated by *Pronuba yuccasella*. Soon after twilight falls these little white moths fly from flower to flower gathering from the anthers with their trunk-like tentacles, which are covered with short spines, the sticky masses of pollen, until a ball, sometimes twice or three times as large as her head, has been accumulated. Then she usually flies away to another plant, and alighting on the seed-pod pierces the wall with her saw-like ovipositor and deposits an egg in one of the rows of ovules. After 3 or 4 eggs have thus been laid the moth ascends to the top of the pistil, and in the funnel formed by the stigmas, which are receptive only on the inner surface, she crowds the ball of pollen. Apparently she intentionally pollinates the flowers, for if she failed to perform this service no seed would be produced and her offspring would perish for want of food. The moth herself receives no direct benefit since her tongue has lost its sucking function and she no longer takes food. (Fig. 67.)

The possession of an ovipositor and spined tentacles by a genus of moths, and the collection and placing of pollen on the stigmas are structures and functions which stand alone in the history of flower-pollination. While it seems incredible that *Pronuba* can understand that unless pollen is placed on the stigmas no seed will be produced, the more her behavior is investigated the stronger becomes the evidence that such is the

143

fact. Riley at first believed that the stigmatic fluid was a form of nectar, which attracted the moth to the stigmas. The ball of pollen, he thought, might have been accumulated accidentally and, proving an encumbrance while the insect was feeding, was dislodged and left in the stigmatic cavity. But twenty years of experience and the discovery that *Pronuba* passed her life without taking food compelled him to admit that her acts were "more unselfish."

Kerner looks upon this act as "unconsciously purposeful," and compares it with the instinct of those caterpillars which, living in the hard parts of wood, before they pass into the chrysalid stage make a special exit by which the tender adult insects may emerge into the world. But it is not difficult to see how this latter habit has arisen. These larvæ undoubtedly at first underwent their transformations outside of the plant, but later remained within their burrows for the sake of the greater protection afforded. Instinctively they continued to make an exit. The habit has never weakened, since any caterpillar which failed to make an opening left no descendants.

Coquillet would regard the behavior of *Pronuba* as a purely intelligent act. "There appears to be no doubt," he says, "that she was in possession of the fact that, unless she did thus pollinate the flower, there would be no seed-pods for her offspring to live on." But it seems past belief that *Pronuba* has long been in possession of knowledge that has only recently become known to the human race. It is far more probable to suppose that the pollen was collected at first for some purpose useful to the species, although it is difficult to imagine in what way, for there is no reason to believe that the larvæ were ever fed with pollen, as in the case of the brood of bees. It will be noticed that the moth gathers the ball of pollen before she lays her eggs, and that, after they have been deposited, she

144

FIG. 67. Twig from Flower-Cluster of *Yucca Whipplei*

1, Bud; 2, flower open; 3, flower open previous night, but now closed; 4, the little white moth
Pronuba yuccasella, flying to the flower; 5, the moth pushing down ball of pollen into the
funnel of stigma; 6, stigma enlarged. (After Kerner)

stuffs it into the infundibuliform stigma. May this not be done for the purpose of protecting her eggs by closing what appears to her like an opening into the seed-capsule? Bees and wasps regularly close the entrances to their burrows for the purpose of protecting their eggs, and it seems possible that the moth began filling the hollow stigma with pollen for the same reason. Assuming that this was the fact, then the flowers thus pollinated would be benefited from the beginning and would produce more seed than flowers depending for pollination on some other agency. The larvæ in the seed-capsules would be assured an abundance of food, and the moths would leave a larger progeny than those not possessing this habit. Once this practice was established, it in time became a necessity. Flowers in the absence of the moths set no seed, and moths failing to perform this service left no offspring. Thus the race of moths pollinating the flowers alone survived.

Many Mexican cacti have large, strongly scented, nocturnal white flowers, which are pollinated by hawk-moths. Among these the night-blooming *Cereus*, or "Queen of the Night," is a not uncommon house-plant in the Northern States, the blooming of which is often chronicled in country newspapers.

Among butterflies and moths the hawk-moths are easily the most important as flower-pollinators, and among anthophilous insects are surpassed only by the bees. There are about 100 species in this country. They are distinguished by their swift, impetuous flight, their large size, their sombre but handsome garb of tan, brown, and gray, sometimes marked with yellow or red, and their elegant forms. They exhibit a high degree of flower fidelity and make their visits with astonishing rapidity, a species of *Macroglossa* having been observed in the Alps to visit several hundred flowers of a primrose in a few minutes. Since the nocturnal species do not fly in stormy weather, but

FIG. 68. A Hawk-Moth. *Sphinx chersis*

mostly on calm, warm evenings and are far from abundant, their swift flight is a great advantage in enabling them to pollinate many flowers in a short time. Hawk-moths have the sense of smell very strongly developed, and consequently nocturnal flowers are usually odoriferous. (Fig. 68.)

Two genera of hawk-moths fly in the daytime, the clear-winged moths *Hemaris* and *Macroglossa*. Some species of *Macroglossa* so closely resemble humming-birds in their appearance and manner of flight that the natives of South America, according to Bates, believe that one can be transmuted into the other. Bates himself, several times by mistake, shot a hawk-moth instead of a humming-bird, and it was long before he could distinguish the one from the other on the wing. The little children of Fritz Mueller came running to him one day and declared in great excitement that they had seen a six-legged humming-bird.

A common cultivated hawk-moth flower is the sweet-scented, climbing honeysuckle (*Lonicera Periclymenum*, Fig. 69.) The flowers expand early in the evening and are at first white within and purplish without. The pistil is bent abruptly downward, while the anthers stand directly in front of the entrance, by which arrangement self-pollination is prevented. The fragrance is very powerful and may be perceived at a long distance. In the daytime Kerner placed a hawk-moth 300 yards away and marked it with cinnabar. When twilight fell, the moth began to move the feelers, which serve it as olfactory organs, hither and thither a few times, then stretched its wings and flew like an arrow through the garden to the honeysuckle. In the dusk I have often seen several species of moths darting swiftly from flower to flower and, as they poised for a few seconds in the air, coming in contact with the anthers and covering the whole under-side of the body with pollen. As a

FIG. 69. Climbing Honeysuckle. *Lonicera Periclymenum*
A hawk-moth flower

landing-stage would be in the way, the lobes of the petals are turned backward. By the second evening the corolla has changed within from white to yellow, the stamens have bent downward, while the stigma has moved upward and now stands in front of the entrance. The flowers are also occasionally visited by bumblebees, which are able to reach a part of the nectar. Very likely the honeysuckle was once a bumblebee flower, but the corolla-tube has lengthened to such an extent in response to the visits of moths that the bumblebees are at present nearly excluded. The flowers are frequently visited by humming-birds in the daytime.

The yellow evening-primrose (*Œnothera biennis*), so common in hedgerows and waste land, is also pollinated by hawk-moths, but in this locality they are not frequent visitors. The flowers expand so quickly at about dusk that the motion of the petals is clearly visible. The anthers are open and are covered with pollen, but the four lobes of the stigma are folded close together. The day following the anthers shrivel and the four stigmatic lobes diverge, forming a cross, which a hawk-moth cannot fail to touch. However it may have been in the past, the flowers at the present time do not appear to attract a sufficient number of visitors, for according to the observations of De Vries in Europe and of Davis in America they are regularly pollinated in the bud. (Fig. 70.)

Close by the evening-primrose I often find the night-flowering catchfly (*Silene noctiflora*), called catchfly because the whole plant is viscid, hairy, and destructive to many small flies. The small white flowers open at sundown, but I have found them very sparingly visited by moths. (Figs. 71, 72, and 73.)

Of the other moth-flowers only a few of the more common can be mentioned here; they are characterized by having white or nearly white flowers which open in the evening, have long

corolla-tubes, and are sweet-scented. In the pink family there
are the sand-pink (*Dianthus arenaria*), bouncing bet (*Saponaria*

FIG. 70. Evening-Primrose. *Œnothera biennis*
A hawk-moth flower

officinalis) common along roadsides (Fig. 74), white lychnis
(*L. album*), and evening-lychnis (*L. vespertina*), and the
long-flowered catchfly (*Silene longiflora*), and the nodding

151

Fig. 71. Night-Flowering Catchfly. *Silene noctiflora*

FIG. 72. White Catchfly. *Silene Armeria*

THE FLOWER AND THE BEE

catchfly (*S. nutans*). Exotic moth-flowers are also common in
the nightshade family, as several cultivated species of tobacco,

the long-flowered to-
bacco (*Nicotiana longi-
flora*), which has a green
tube 4 inches long, and
the similarly flowered
night-blooming tobacco
(*N. noctiflora*, Fig. 74A),
and the white flowers
of the Jamestown weed,
which are 3 inches long.
Two species of lilies are
pollinated by hawk-
moths, as *Lilium candi-
dum* and *L. Martagon*,
as are also several spe-
cies of gentians (*Gen-
tiana verna* and *G.
bavarica*), the vernal
crocus (*Crocus vernus*),
several kinds of the
sweet-scented *Gardenia*
and a number of or-
chids in the genus *Ha-
benaria*. (Fig. 75.)

Several years ago at
Cambridge, Mass., I
saw in bloom that re-
markable orchid from

FIG. 73. Racemed Catchfly. *Silene
dichotoma*

Madagascar, *Angræcum sesquipedale*, which bears large, snow-
white flowers with a slender green nectary of the astonishing

154

FIG. 74. Bouncing Bet. *Saponaria officinalis*
A hawk-moth flower

FIG. 74A. Night-Blooming Tobacco. *Nicotiana noctiflora*
A hawk-moth flower

Fig. 75. White Variety of Purple-Fringed Orchis. *Habenaria psychodes*

length of 11 inches. It seemed impossible at the time of the discovery of this plant that there should be in existence a moth with a tongue long enough to consume all of the nectar, but such a moth was later actually found. To remove the pollen masses it must thrust its long proboscis into the nectary up to its very base. If these great moths were to become extinct, then assuredly *Angræcum* would also become extinct, for smaller moths are unable to remove the pollinia. On the other hand, as Darwin states, there will always be an inch or more of nectar at the base of these long nectaries safe from the depredations of other insects, upon which the moths are probably largely dependent. Thus the destruction of either the plant or the moth would be fatal to the survivor. The tongues of the moths tend continually to increase in length in order that they may drain the last drop of nectar, while the nectaries which are long enough to compel the moths to insert their tongues up to the very base will be the best fertilized. Thus there is a race between the moths and the plants, in which *Angræcum* has triumphed, for it still flourishes abundantly in the forests of Madagascar.

There was undoubtedly a time in the history of butterfly and moth flowers when the nectar was less deeply concealed than at present and was accessible to bees. Concealment of the nectar was at first beneficial by shutting out marauding beetles and flies. And at this point the lengthening of the nectaries, or corolla-tubes, should have stopped, for bees are the most valuable of pollinators, and as a general principle it is a disadvantage for a flower to be dependent on a single species or genus of insects for pollination. But variation in the direction of increased length of the nectaries once started, the impulse still continued, and the tongues of the visiting moths and butterflies lengthened correspondingly. The necessity of building nests and caring for their young made this impossible in the case of bees. Just as the momentum of a swinging pendulum carries it beyond the central point of equilibrium, so the momentum of the variation carried the length of the nectaries to a point where bees were excluded. Thus butterfly and moth flowers came into existence; but this reciprocal dependence does not in most cases imply more effective pollination, although the Lepidoptera are assured a larger supply of nectar. Tendency to vary in a definite direction, even when no benefit is derived, is shown by the increasing complexity of the markings on the wings of insects and the convolutions on the margins of shells. It is not at all improbable that the nectaries of certain flowers and the tongues of the visiting Lepidoptera may continue to lengthen until both become extinct.

An intricate mechanism to effect pollination does not prove that such an arrangement is best for flowers in general. It finds its explanation in the particular conditions under which each flower was developed. Nature seems at times to be a very poor teleologist. The wonderful orchids are less successful than many lowly dooryard weeds. In the case of certain flowers orthogenesis, or determinate variation in a definite direction, has carried forward their specialization until they are face to face with extinction.

CHAPTER X

FLY–FLOWERS

THAT the physical characters of flower-visiting insects, such as size and the length of the tongue, should influence the structure of flowers would be expected; but it is more surprising to find their mental traits also reflected. How different is the reception accorded by flowers to many stupid flies from that given to bees! Notice how the constant and observant bees are offered nectar, pollen, shelter, an alighting-platform, bright hues, and sweet odors, while undesirable guests are excluded. But for the unspecialized, stupid flies there are pitfall-flowers, prison-flowers, pinch-trap flowers, and flowers with deceptive nectaries, deceptive colors and odors. In her readiness to take advantage of their weakness Nature simulates the worse qualities of humanity, although, more strictly speaking, it is their inability to learn from observation that has induced the development of these peculiar forms. But not all flies are stupid. This is far from true of the syrphid or hover flies and the bee-flies which visit nearly the same flowers as do the bees.

Their numbers and activity probably entitle the flies to rank after the bees and before the butterflies and moths as flower-pollinators. Mueller places them next to the Hymenoptera, except in the case of the Alpine flora, where butterflies are very abundant. In New Zealand both bees and butterflies are very scarce, and Thomson considers the flies as the chief agents in pollination.

There are in North America more than 8,000 described species of two-winged flies, or Diptera. Very many of them live largely or wholly on animal substances, and never, or only rarely, visit flowers. To this group belong mosquito-like flies with long antennæ, small heads and eyes, slender bodies, and long legs, as the crane-flies, midges, mosquitoes, punkies, gall-gnats, March-flies, and black-flies. Flies which visit flowers frequently for nectar and pollen resemble the house-fly, and usually have short antennæ, large heads and eyes, robust bodies, and short legs, as the horse-flies, soldier-flies, robber-flies, bee-flies, house-flies, dance-flies, syrphid flies, and flesh-flies.

The habit of visiting flowers has been acquired independently in many different families of flies, and all the intermediate stages may be found between forms which are predaceous and those which live wholly on floral food. Mosquitoes, especially the males, occasionally visit flowers, and one genus (*Megarrhinus*) never sucks blood, but in both hemispheres has been observed to feed on nectar alone. The males of the horse-flies, or blood-thirsty *Tebanidæ*, live on nectar, while the females usually suck the blood of animals, but occasionally visit flowers. The syrphid flies in the adult stage depend chiefly on pollen and nectar, while the bee-flies feed only on nectar.

Flowers with nauseous or indoloid odors, due to the decomposition of some nitrogenous compound, are attractive to flesh or carrion flies. The petals are often flesh-colored, blood-red, dull dark-purple, marked with lurid stripes or spots. To some observers they suggest putrefying flesh or decaying carcasses, but in most instances the resemblance is not very apparent. There are also malodorous flowers which are yellowish green or white. It is chiefly the nauseous odor rather than a likeness to putrid substances which draws to flowers carrion and dung flies belonging to the genera *Musca*, *Lucilia*, *Calliphora*, and *Sarcophaga*. Many strong-scented odors, which are not repulsive, are also attractive to flies.

In damp thickets on the banks of rivers, twining amid the
bushes there grows a pretty vine with smilax-like leaves and
umbels of green flowers pollinated by flies. (Fig. 76.) This

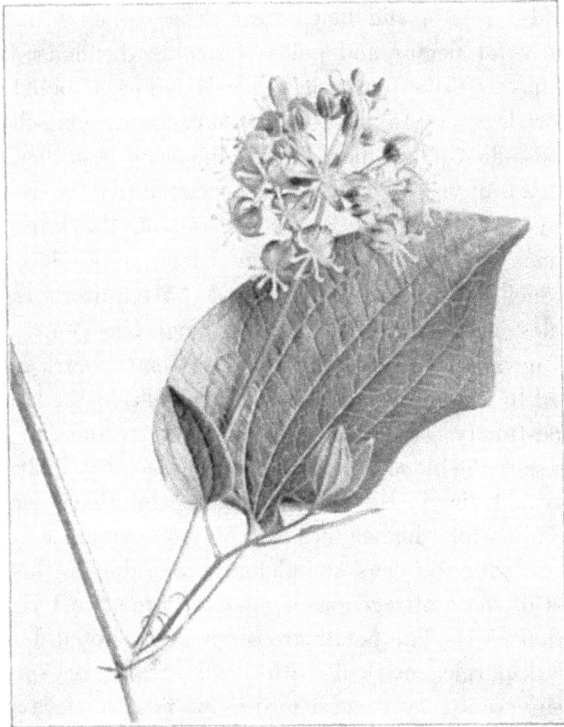

FIG. 76. Carrion-Flower. *Smilax herbacea*
A fly-flower

is the carrion-flower and so offensive is its odor that it well
merits the name. Another carrion-flower is the purple trillium
(*Trillium erectum*). (Fig. 77.) In early spring children often
gather bouquets of its lurid, purple flowers which they are soon

Fig. 77. Purple Trillium. *Trillium erectum*
A fly-flower

compelled to throw away. Since the blossoms are nectarless, they are visited only occasionally by flies for pollen. Its penetrating characteristic odor, it is needless to say, prevents any one from making such a mistake in the case of the skunk-cabbage. Within each spathe or leafy hood there is a cluster of small,

163

perfect flowers attractive to flies. The disagreeable odor might be supposed to be repellant to bees, but sometimes in early spring large numbers of them gather pollen from this source for brood-rearing. The water-arum (*Calla palustris*), a plant growing in cold bogs, has a handsome white spathe, but the nauseous scent places it among fly-flowers. (Fig. 78.)

Among flowers with unpleasant odors the saxifrages are highly interesting, not because flies are the most numerous visitors, but because the white corolla is covered with many-colored dots. The white petals of the round-leaved saxifrage (*Saxifraga rotundifolia*) are sprinkled with round dots, the outer of which are intense purple red and the inner yellow. The snow-white flower of the star saxifrage (*S. stellaria*) are beset with purple dots and adorned with two orange-yellow spots. *Saxifraga bryoides* is white with many shining yellow dots. (Fig. 79.) The large golden-yellow flowers of *S. aizoides* are marked with numerous orange-red dots, and are visited by 85 different species of flies; all of these spotted flowers are indeed very frequently visited by flies. Mueller believed that the dots attracted the attention of flies more than of other insects.

There is another group of fly-flowers, the pitfall-flowers, which rely partly on deceptive odors and colors, and partly on pitfalls, which are veritable prison-traps—often death-traps—to various unwary flies. The spotted arum (*Arum maculatum*) of Europe is a prison-flower and, like all the Aroids, has an offensive odor. The spathe, which is broad above, is constricted in the middle into a neck, below which there is a bulbous cavity completely ensheathing the lower part of the club-shaped flower-stalk, or spadix. In the neck or entrance to the chamber there is on the spadix a ring of bristles with the points inclined downward, a little below this a band of staminate flowers, then another ring of hairs, beneath which are the pistillate flowers.

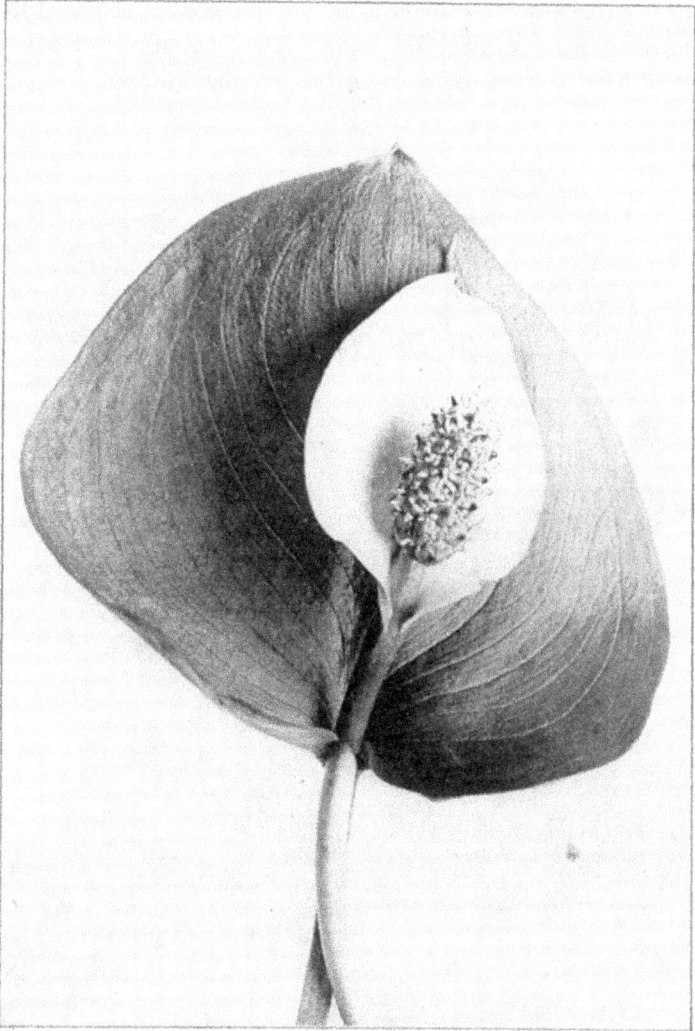

FIG. 78. Water-Arum. *Calla palustris*
Exhales a disagreeable odor attractive to flies. Grows in wet swamps

The pistillate flowers which secrete nectar mature first, and the incurved hairs permit the midges (*Ceratopogon*) which effect pollination to pass easily down to them, but prevent their re-

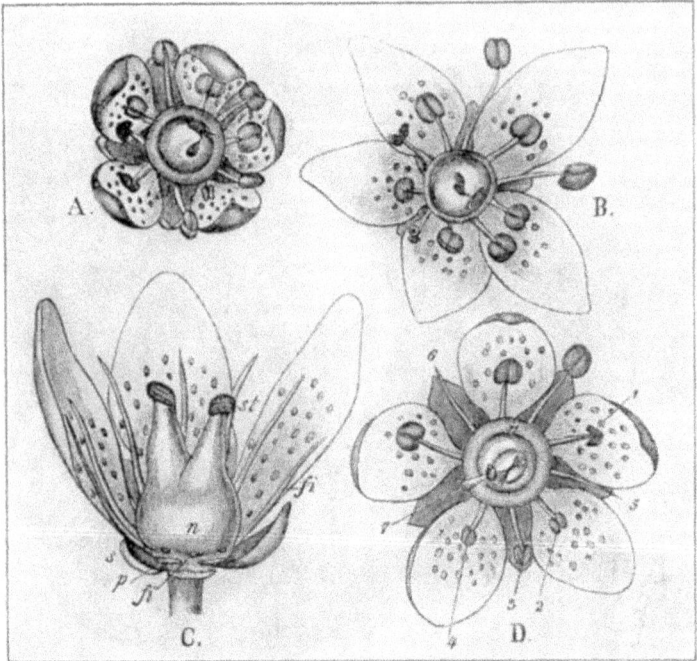

FIG. 79. Mountain Saxifrage. *Saxifraga bryoides.* (After Mueller)

The petals are covered with numerous dots or spots, which Mueller believed to be attractive to flies, which visit the flowers in large numbers. *A*, *B*, and *D* represent the flower in the first stage, when the anthers mature, but the stigmas remain unreceptive; *C* represents the second stage, when the stigmas are mature and the anthers have withered and fallen away

turn. The lowest ring of hairs withers first, permitting the midges to come up to the staminate flowers, where they become covered with pollen, then the upper ring of hairs withers and the midges are free to fly away. As many as a thousand midges may be imprisoned in a single spathe. (Fig. 80.)

To the same arum fam-
ily, or *Araceæ*, belongs
jack-in-the-pulpit, also
called Indian turnip from
the shape of its root (*Arisæma triphyllum*). It is a
pitfall-flower. It flourishes
in wet swamps and my
observations were made
while standing up to my
ankles in water and sur-
rounded by a cloud of mosquitoes, from which a veil
gave me protection. The
staminate and pistillate
flowers are on different
plants. The staminate are
much the smaller, being
only 6 or 7 inches tall,
bloom first, and soon pe-
rish. The pistillate or fertile plants are much larger,
often 2 feet tall. The
spathe is dark purple
striped longitudinally with
white, and ensheathes a
club-shaped stalk or spadix,

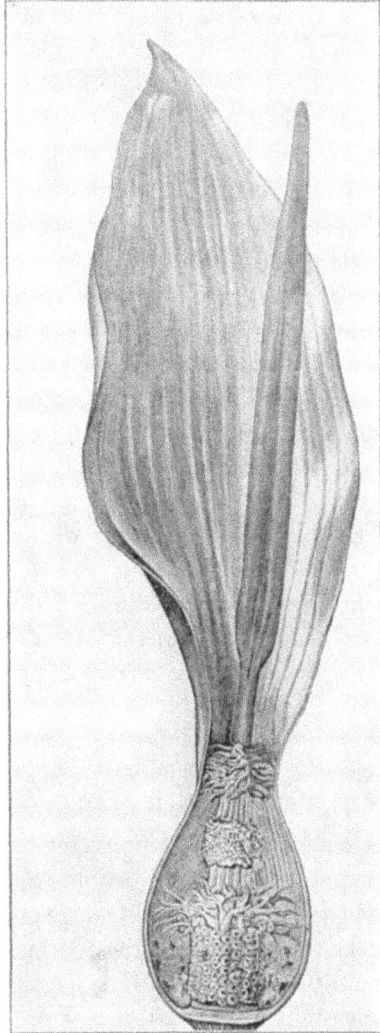

Fig. 80. Arum.
Arum conocephaloides

A prison-flower. To show the arrangement of the small flowers the front of
the spathe is removed. On the low-
est part of the spadix, or club, are the pistillate ("female") flowers, above them the first
ring of bristles, next the staminate ("male") flowers, and then a second ring of bristles.
At the bottom of the cavity are a number of midges (*Ceratopogon*), whose escape is prevented by the stiff reflexed bristles of the lower ring. (After Kerner)

which bears near its base several whorls of small, naked flowers, while its apical end is arched over the "pulpit" to exclude rain. There is a small orifice at the base of the spathe, where one edge overlaps the other, which serves for the purpose of drainage. (Figs. 81 and 82.)

The inner side of the spathe is smooth, shining, and very slippery, far more highly polished than the outside. When little moth-like flies of the genus *Psychoda* rest on this polished surface they are unable to gain a foothold and fall into the chamber below. There for a time they are held prisoners, since they cannot climb the smooth walls or the equally smooth base of the spadix. The staminate flowers are visited first, since they bloom first. As the spathes wither, their inner surfaces relax and become rougher, enabling the little visitors, now loaded with pollen, to escape and fly to the pistillate plants. The spathes of the latter wither less promptly, but from the point of view of the "jacks" this is of little consequence, since pollination has been effected; but it is fatal to many of the flies which, unable to escape, perish in the chamber. The arum family includes many tropical forms, like the calla-lily; about the pollination of most of these very little is known and undoubtedly many remarkable facts await discovery.

The peculiar-shaped Dutchman's-pipe (*Aristolochia sipho*) is pollinated in a manner very similar to that of jack-in-the-pulpit; but the flowers are perfect, *i. e.*, contain both stamens and pistils. The calyx hangs downward, is about an inch and a half long, bent like the letter S, constricted in the middle, with the bowl-end of the pipe narrowed at the throat and very smooth within. After they have once entered this tubular passageway small flies are unable to fly or creep out until the calyx withers. (Fig. 83.)

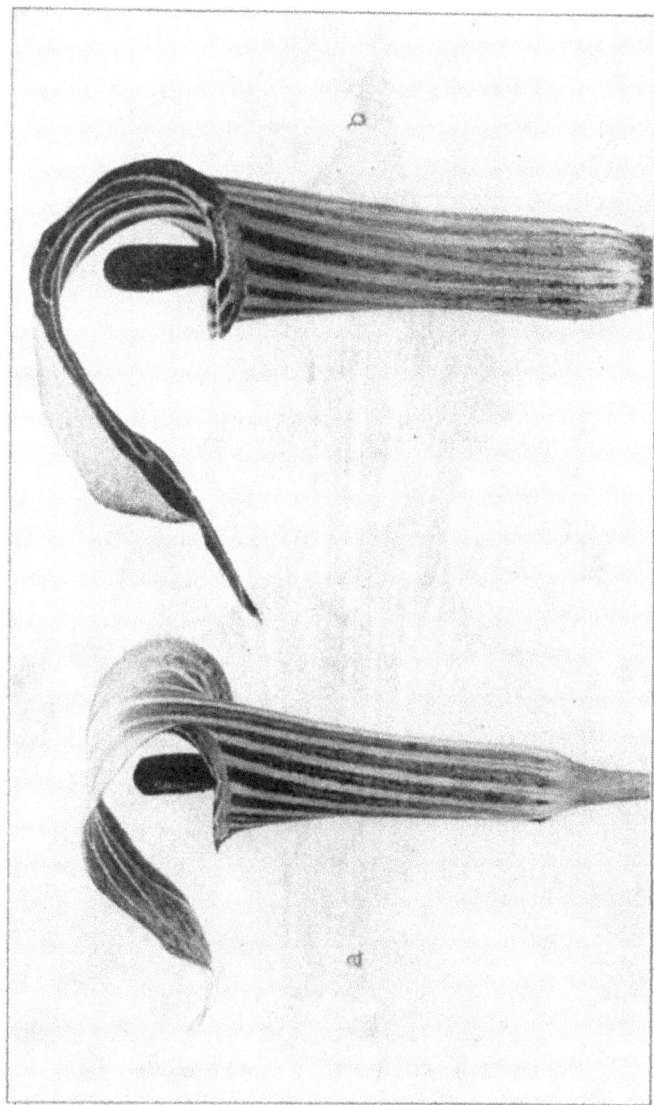

Fig. 81. Jack-in-the-Pulpit. *Arisæma triphyllum*
A, staminate flowers; *B*, pistillate flowers. A fly-flower

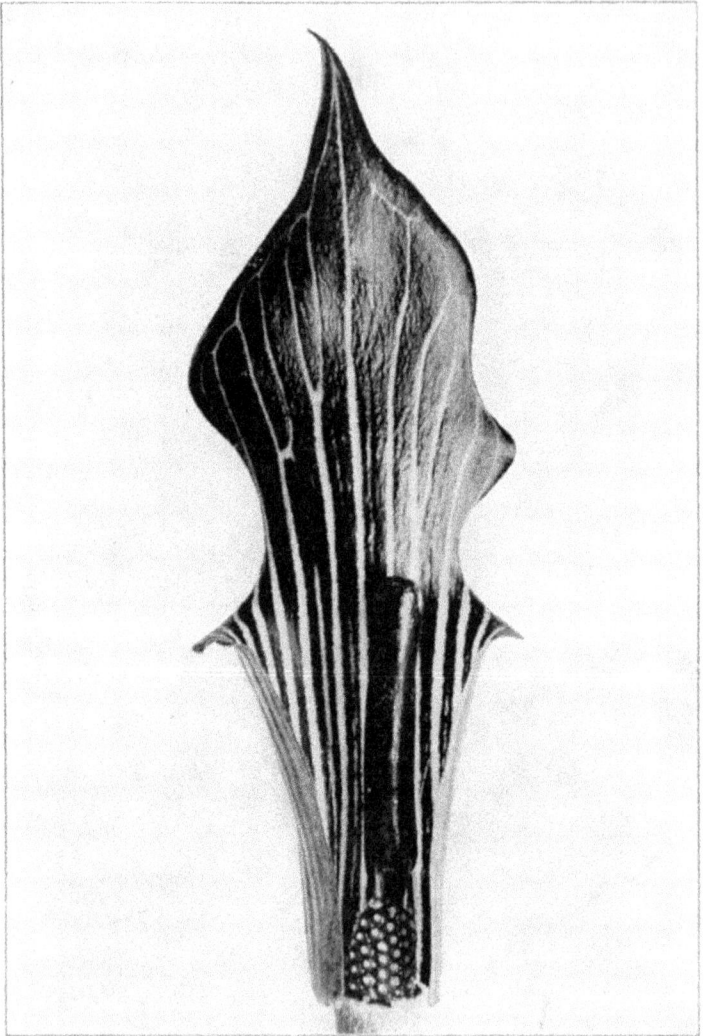

FIG. 82. Jack-in-the-Pulpit. *Arisæma triphyllum*

Spathe or ensheathing leaf of pistillate flowers opened, showing fertile flowers at base of spadix

Among the Diptera the family which most frequently visits flowers and is of the most importance in pollination are the

FIG. 83. Dutchman's-Pipe. *Aristolochia Sipho*
A fly-flower

hover-flies or *Syrphidæ*. They feed on both pollen and nectar, and are found on many different species, their long tongues enabling them to reach the nectar in many bee-flowers. There are several small flowers adapted to pollination by the hover-

171

flies, the most common being the speedwells, tender little herbs of the genus *Veronica*, which grow in our gardens, lawns, and meadows. When June is a wet month the thyme-leaved speed-well (*Veronica serpyllifolia*) is abundant. The white or pale-blue petals are marked with deeper purple lines leading to the nectar; the corolla-tube is yellow and the throat is fringed with

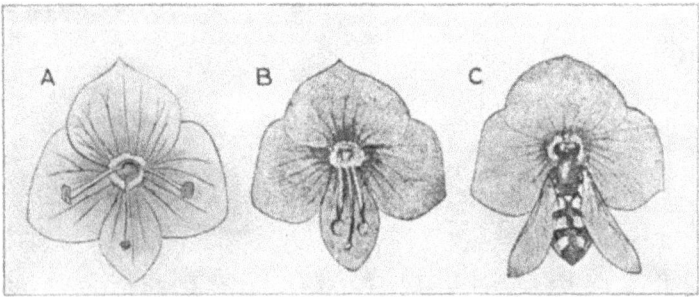

Fig. 84. Speedwell. *Veronica Chamædrys*

A syrphid-fly flower. *A*, flower seen from the front; *B*, the same with the stamens placed together; *C*, the same with a syrphid fly sucking nectar. (After Mueller)

hairs to exclude water. (Fig. 84.) There are only two sta-mens, one on each side of the flower, and a single pistil in the centre, all three of which taper at the base so that they bend easily. When a syrphid fly visits one of these flowers, the stigma rests against the under-side of its body while the feet grasp the stamens and draw them also beneath it, where they leave a part of the pollen. Self-pollination may be prevented in some species by the stigma maturing before the anthers, but in the thyme-leaved speedwell the anthers may deposit pollen directly on the stigma.

The syrphid flies are very common visitors to golden-yellow flowers, like the buttercup and marsh-marigold, and Mueller, observing that they often poised before them a few seconds before darting down to suck nectar or eat pollen, thought that the bright-yellow coloration must afford these flies an æsthetic pleasure. But more recently Plateau has shown that this is merely a habit of flight, and that the hover-flies poise on the wing in the same way before green leaves, green flowers, green fruits, and green and brown stems, or even before a cane, or a marble, or the human finger. When a finger was gently interposed between a fly and a flower, the fly remained in a fixed attitude as before and, when the finger was slowly moved away, followed it.

Many syrphid flies closely resemble bees and wasps, and can be distinguished from them with difficulty when on the wing. Field-collectors occasionally catch them and send them away as wasps. By some naturalists this similarity of form and flight is regarded as a case of protective resemblance, by others as an accidental convergence in likeness.*

No one can long observe flowers without meeting with the bee-flies (*Bombyliidæ*), which both when flying and at rest are often mistaken for bees. They feed on nectar, but not on pollen. The species of *Bombylius* are densely clothed with silky hair and both in movement and appearance suggest small moths. The long proboscis points directly forward, and

* An amusing illustration recently occurred in the experience of the writer showing how easily certain flies may be mistaken for honey-bees. A prominent local official, accustomed to observe carefully, told me how his wife had called his attention to the presence of many bees on a window of a shed chamber. He related how after putting on an overcoat and protecting his face and hands, he had finally driven them outdoors.

"Now," he inquired, "how did they get there?"

"They were not bees at all," I replied, "but flies. If you will examine them carefully you will find that they have only one pair of wings."

Naturally he was much astonished at this statement; but some days later he brought me two of the insects in a bottle. They proved to be, as I had expected, syrphid flies belonging to the species *Eristalis tenax,* often found on flowers.

the flies dart swiftly about and poise on the wing when sucking, in the same manner as the hawk-moths.

In the Alps bee-flies were observed by Mueller to visit three times as many red, violet, and blue flowers as yellow and white ones. But in Wisconsin Graenicher's observations show nearly the opposite result, "in other words, more than twice as many white and yellow flowers have received the attention of these flies as red, purple, or blue ones." In reply to an inquiry of the writer's as to how he accounts for Mueller's results, he replies that it may be explained by a larger number of red and blue flowers in the Swiss flora than in that of Wisconsin. He also calls attention to the curious fact that during seven years in Germany *Bombylius discolor* was observed to prefer the flowers of the lungwort (*Pulmonaria officinalis*) to those of any other plant, visiting other flowers only in the case of necessity.

The dance-flies (*Empididæ*), which may often be seen near running brooks dancing in mid-air beneath the foliage of the trees, have long tongues, and resort to many flowers for nectar. I shall long remember the fair June afternoon spent in collecting the visitors of the twinflower (*Linnæa borealis*). In an open grove of tall hemlocks a large bed of this trailing evergreen vine was pink with nodding blossoms. (Fig. 85.) The flowers, which exhale a sweet, vanilla-like fragrance, are borne in pairs at the summits of elongated peduncles. The inverted position of the flowers excludes the rain. The nectar is secreted at the base of the corolla, on the lower side of which there is a yellow marking or honey-guide. Within the corolla are many inter-crossing hairs which shut out small, useless flies, which I have seen vainly seeking an entrance. During the afternoon I collected eight visitors, all of which were found on examination to belong to a single species of dance-fly (*Empis rufescens*). Other observations show that this fly is the only pollinator in

Fig. 85. Twinflower. *Linnæa borealis*
A fly-flower

FIG. 86. Common Milkweed. *Asclepias syriaca*

A pinch-trap flower

this locality, and that the twinflower is an *Empis* flower. The males of the dance-flies live wholly on flowers, but the females are partly predaceous.

The milkweeds are pinch-trap flowers, which in their remarkable pollinating mechanism rival the orchids. The pollen coheres in waxy masses called pollinia, which by means of an ingenious clip-mechanism are clamped to the legs, tongues, and antennæ of flies and many other insects. When the insect flies to another flower the pollinia come in contact with the stigma, to which they adhere so firmly that it can only obtain its liberty by snapping the connecting bands. Only a part of the species of *Asclepias* are pollinated by flies, others are pollinated by bees and butterflies. It is not uncommon in examining specimens of bees to find one or more of these clips, which are a useless burden and interfere with their work, on their legs and antennæ. Gibson states that an English bee-keeper lost thousands of his bees from the effects of strings of these clips, which it was at first thought were a fungous growth. (Fig. 86.)

If an insect caught in one of these pinch-traps is not strong enough to pull away the pollinia or break the bands, it is held a prisoner and dies a lingering death, although probably a nearly painless one. In New Zealand the flowers of *Araugia albens*, a plant introduced from the Cape of Good Hope, which are normally pollinated by bumblebees, catch in a single night hundreds of moths. It was once seriously proposed by an economic entomologist to employ this plant in the extermination of the codling-moth, so injurious to apples; but unfortunately the co-operation of the codling-moth could not be obtained, for it persistently refused to visit the flowers of *Araugia* —another illustration that the well-laid schemes of mice and men—

"Gang aft a-gley."

CHAPTER XI

BEETLES AND FLOWERS

THERE are no beetle-flowers, although Delpino believed that the flowers of magnolia were adapted to flower-beetles of the genus *Cetonia*. It would be much better, indeed, for flowers if they were never visited by beetles, for they accomplish far more harm than benefit. As agents in pollination they are of little significance. The enormous devastation of foliage and bloom, the absence of hairs for holding pollen, the consumption and waste of pollen and nectar, the inactivity of many species, and their indefinite manner of flight are factors which greatly reduce their value as pollen-carriers. As a whole they are highly destructive to vegetation and cause annually much loss to the farmer, fruit-grower, and florist. The cherry-weevil often destroys the crop of plums and cherries, the rose-chafer strips the rose-bushes of both flowers and foliage, while the blister-beetles may devour large areas of tomatoes and potatoes in a few days.

But the order of Coleoptera, or beetles, is of special interest, as Mueller has pointed out, because it shows so clearly how the habit of anthophily, or flower-visiting, has arisen, and its beginnings in many different families, genera, and species. The length of time which has elapsed since the first tendency in a family toward this habit was manifested is indicated by the number of species which resort to flowers. Where a whole family is dependent on a floral diet the epoch was more remote, but where there are only a few isolated species the habit has been acquired more recently. In the most diverse families

Fig. 87. Beetles Which Never Visit Flowers

1, Stag-beetle (*Lucanus cervus*); 1*A*, male; 1*B*, female; 2, horned beetle (*Oryctus nasicornis*); 3, tiger-beetle (*Cicindela 6-guttata*); 4, *Prionus laticollis*; 5, goldsmith-beetle (*Catalpa lanigera*)

single species have become habituated to a diet of pollen and nectar, and subsequently in some cases structural changes have been developed to insure greater success in the search for food.

In New England Frost and myself have collected 232 species of beetles, belonging to 127 genera and 29 families which visit flowers. This is probably less than $\frac{1}{10}$ of the total number of described species in this region. In all Europe Knuth has enumerated only 434 anthophilous beetles. Why is it that so few feed on pollen and nectar? Their habits and forms, in many instances, answer this question; many are predaceous like the tiger-beetles and ground-beetles, or are scavengers like the rove-beetles; others are nocturnal, or aquatic, or occur chiefly on the ground, lurking beneath stones and boards, or living in the nests of other insects. (Fig. 87.) Many flowers, moreover, have the nectar concealed where it is beyond the reach of beetles, which, with few exceptions, have short tongues.

Beetles are usually found on common, open flowers with the nectar visible or nearly visible, as the cherries, cornels, shadbush, New Jersey tea, and goldenrods. In early spring, on the white flowers of the shadbush (*Amelanchier canadensis*), 31 species have been taken, the choke-cherry (*Prunus virginiana*) (Fig. 88) yielded 43, the chokeberry (*Pyrus arbutifolia*) 10, and the dense panicles of small white flowers of the meadow-sweet (*Spiræa salicifolia*) 42, while the flat-topped cymes of the cornels (*Cornus*) attracted 38 species, and the profusion of flower-clusters, which convert the bushes into huge bouquets, drew to the viburnums (Fig. 89) the phenomenal number of 81 kinds. In New England on the New Jersey tea (*Ceanothus americanus*) 13 species were captured, but in the warmer climate of Virginia, Banks's list of beetles taken on the *Ceanothus* includes 58 kinds. Among the *Compositæ* beetles are more common on the flowers of the goldenrod than any other genus;

FIG. 88.   Choke-cherry.   *Prunus virginiana*

The flowers are great favorites with beetles

both pollen and nectar are abundant and the nectar is concealed in floral tubes only $1\frac{1}{2}$ mm. long, while the temperature of the dense clusters, which offer excellent hiding-places, is usually a little above that of the atmosphere, especially at night. The asters, thistles, and sunflowers have longer floral tubes and are visited much less often by beetles. There are a great many flowers, as in the mustard and carrot families, on which only one or two species of Coleoptera have been taken.

In the flowers enumerated nectar is a more important allurement than pollen, which is not produced in large quantities. But many flowers which are totally devoid of nectar are visited by beetles for pollen. The staminate cones of the Scotch pine (*Pinus sylvestris*) and of the fir attract many beetles, while many may also be swept from grasses where they devour the anthers as well as the pollen. Beetles are likewise common on conspicuous pollen-flowers like the rose, poppy, and St.-John's-wort, and at times entirely strip the rose-bushes of both flowers and leaves. The frequency with which beetles resort to pollen-flowers led Knuth to conclude that they prefer pollen to nectar, but if this were universally true, genera feeding on nectar alone, such as *Nemognatha*, would never have been developed.

Beetles are occasionally taken on two-lipped, or bilabiate, flowers with the nectar deeply concealed, where they are either searching for food or their presence is accidental. Although the nectar is inaccessible, it is often possible for them to obtain pollen. According to Kerner many small beetles find a refuge in the interior of gentian-flowers, while species of *Cetonia* remain for several days in the partially expanded flowers of magnolia feeding on the sweet juices and pollen. Carrion-beetles are sometimes found in great numbers in the ill-smelling spathes of the Aroids, and tubular flowers often provide nocturnal lodging for wayfarers of the Coleoptera.

FIG. 89. Sheepberry. *Viburnum Lentago*
Visited by more species of beetles than any other New England flower

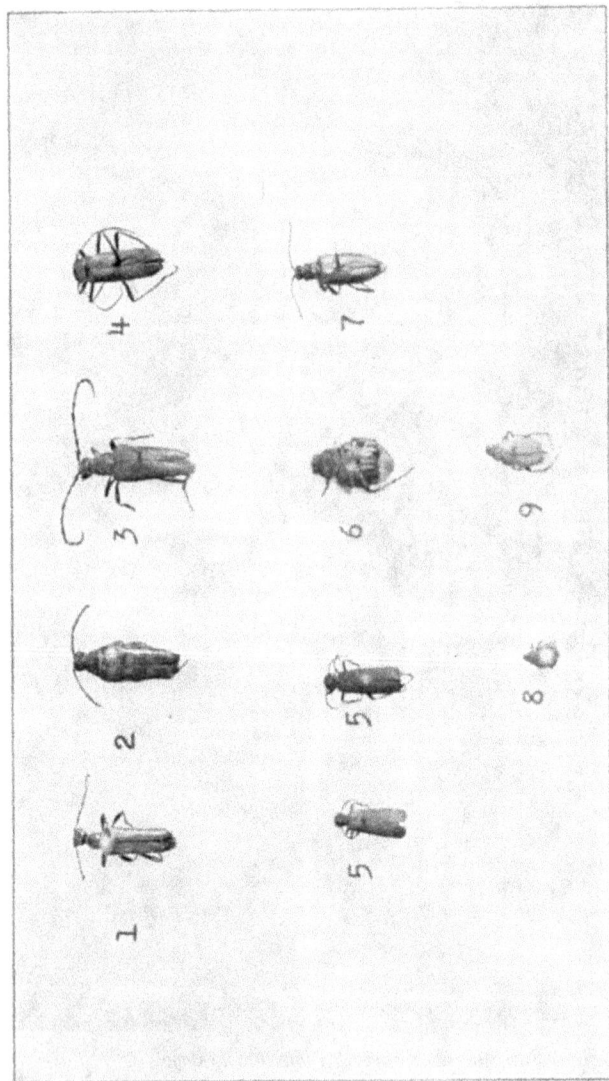

FIG. 90.   Common Flower-Visiting Beetles

Long-horned beetles: 1, *Leptura vittata*; 2, *Typocerus velutinus*; 3, *Leptura canadensis*; 4, soldier-beetle (*Chauliognathus pennsylvanicus*); 5, blister-beetle (*Epicauta pennsylvanica*); 6, hairy beetle (*Trichius affinis*); 7, water-lily beetle (*Donacia piscatrix*); 8, blue-flag beetle (*Mononychus vulpeculus*); 9, rose-chafer (*Macrodactylus subspinosus*)

BEETLES AND FLOWERS

Many beetles pass their entire life on a single plant species. The larvæ of *Donacia piscatrix* mine in the leaves and stems of the yellow water-lily, while the adult beetles flourish within the floating flowers; another species of *Donacia* attaches its cocoons to the base of the stems of the marsh-marigold, and when the flowers open they emerge and climb the stems and live in plenty, half buried among the stamens; the larvæ of the familiar asparagus-beetle eats the leaves of the cultivated asparagus, and the beetles visit the flowers. (Fig. 90, No. 7.)

The blue-flag beetle (*Mononychus vulpeculus*) passes its entire life on the blue flag, and is most common during the blooming-time of the flowers. (Fig. 91.) It is inactive in the bright sunshine, says Needham, and will dodge around the base of a flower like a squirrel around the base of a branch when a hand approaches, but will rarely fly. "With its beak it sinks a shaft in the nectariferous tissue, nibbles a little, makes another hole, and another, and another, until the nectar is left flowing from many punctures, attracting swarms of insects of all sorts." In one instance while the weevil was gnawing a hole, there were three flies facing it and another on its back, "crowding one another like pigs around a trough." The eggs are laid in the seed-capsule, the larvæ feed on the young ovules until they undergo their transformation into beetles, and finally in the fall the bursting of the capsule sets free both the weevils and the seeds. (Fig. 90, No. 8.)

The blister-beetles also restrict their visits chiefly to one kind of flower; for instance, an oblong, dull-black species (*Epicauta pennsylvanica*) is much more commonly found on the flowers of the goldenrod than elsewhere. At times the blister-beetles appear suddenly by bushels and destroy in a few days large patches of potatoes and tomatoes. The larvæ are brood-parasites on bees, grasshoppers, and other insects.

When they first hatch they are active, louse-like forms called triangulins because each leg terminates in three claws. The eggs are laid on the ground near the stem of a flowering plant, and as soon as the triangulins are out of the egg they climb to the flowers, where they wait for the arrival of some insect. (Fig. 90, No. 5.)

Unfortunately for them, they are unable to recognize their hosts, and jump aboard the first conveyance that comes along, whether it is a bee or a fly, with the result that they are often carried far away from the nests they are seeking to reach. There is nothing for them to do but to keep on trying until they either die from exhaustion or, by a happy chance, lay hold of the right insect. Hundreds do perish, and to compensate for this loss the female lays some 2,000 eggs. If, however, a triangulin is carried to the nest of a host bee it feeds on the pollen until it is transformed into a beetle. The adventures of a triangulin are analogous to those of a grain of pollen. Wasteful as is this method, it succeeds much better than would seem possible.

A part of the Coleoptera are sarcophagous, or flesh-eaters, and a part are plant-eaters, or phytophagous, feeding on wood, sap, leaves, and other vegetable matter. The first group is certainly older and more primitive than the second, while among plant-eating beetles those living on wood (xylophagous) are older than those feeding on foliage or flowers. Beetles living on pollen and nectar are the most recent in origin of all.

THE SARCOPHAGOUS BEETLES AS FLOWER-VISITORS

Carnivorous families of beetles, especially where they live on the ground, are not likely to visit flowers. None of the terrestrial tiger-beetles or water-tigers, both of which are

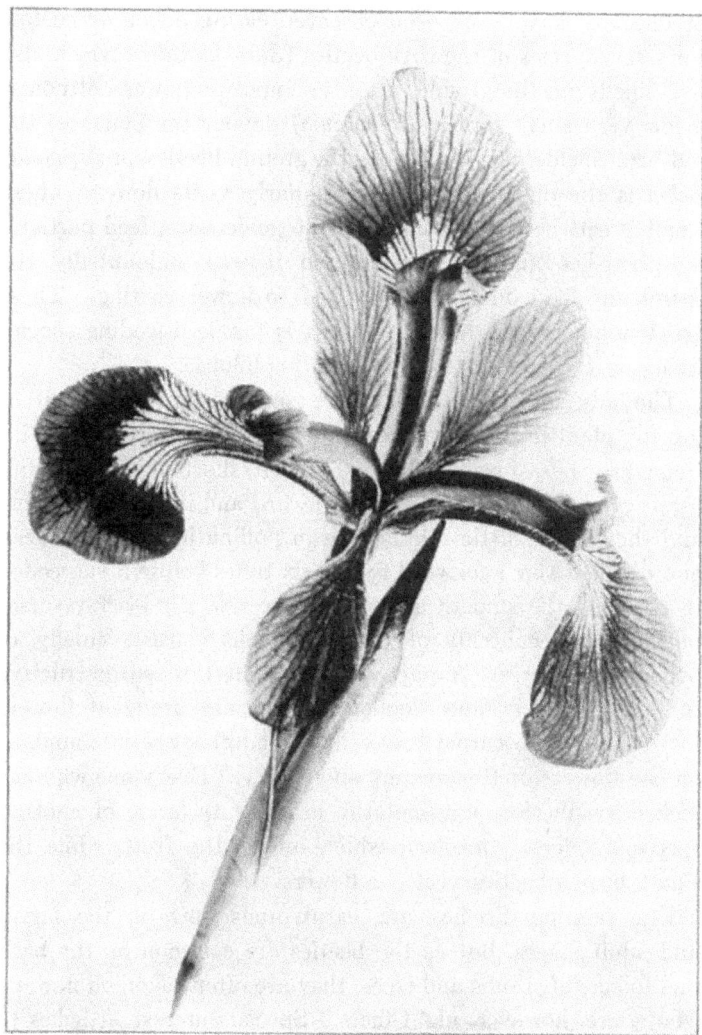

Fig. 91. Blue Flag. *Iris versicolor*

predaceous, have ever been observed eating pollen or nectar. Very few species of the rove-beetles (*Staphylinidœ*), which also live chiefly on the ground, are ever found on flowers, although a few very small forms (*Anthobium*) devour the pollen of the red-berried elderberry. Among the ground-beetles, or *Carabidœ*, *Lebia* is the only genus which regularly visits flowers; these small green beetles, common on the goldenrods, feed partially on plant-lice and insect-eggs, and it was undoubtedly the search for food on foliage that led to flower-visiting. These great families show how difficult it is for flesh-feeding species living on the ground to become anthophilous.

The lady-bugs (*Coccinellidœ*) are common on foliage, searching for plant-lice, or Aphides, and consequently they not infrequently pass over to flowers. Their short legs and round forms render them exceedingly awkward and inefficient visitors and they are of little significance in pollination. I have seen one of them slip backward five or six times before it succeeded in climbing the smooth stem of a flower of the prickly sarsaparilla. The contents of their stomachs consists chiefly of pollen and spores. A part of the dermestids, so destructive to skins, feathers, and woollen carpets, also frequent flowers for pollen. The carpet-beetle in the adult stage is abundant on the flowers of the currant and cherry. Every one who has picked raspberries has met the little white larvæ of another species (*Byturus unicolor*), which infests the fruit, while the small, brown beetles visit the flowers.

The familiar fireflies are carnivorous both in the larval and adult stages, but as the beetles are common on the bark and foliage of shrubs and trees, they are often taken on flowers. Many are, however, night-fliers. Special interest attaches to the soldier-beetles (*Chauliognathus*), one of the few genera of the Coleoptera which have the mouth-parts lengthened to enable them to suck nectar more easily. The soldier-beetle may be found by thousands on the goldenrods, New Jersey tea, linden, and wild hydrangea. (Fig. 90, No. 4.)

It is evident that carnivorous beetles which seek their prey on vegetation are much more likely to acquire the habit of visiting flowers than those which live wholly on the ground. In general, they feed more freely on pollen than on nectar, partly because it is more easily obtained, and partly, perhaps, because it resembles in its composition the animal food to which they are accustomed. Most, if not all of them, have acquired the habit of visiting flowers independently of each other.

THE PHYTOPHAGOUS BEETLES AS FLOWER-VISITORS

But it is among the beetle families, which both in the larval and adult stage feed on vegetable substances, that the habit of visiting flowers has become most important. It is an easy step for them to learn to live on floral food, although from a great number of tubular flowers they are largely excluded.

The click-beetles (*Elateridæ*), of which 39 species have been taken on flowers in New England, live under bark, or bask in the sunshine on the foliage of trees and herbage. The leaf-beetles (*Chrysomelidæ*), one of the latest families to be evolved, are of small or medium size and in both the larval and adult stages are very destructive to foliage. It is only by the expenditure of much time and labor that the ravages of the potato-bug and squash-bug are checked annually. It is impossible for this immense family to depend on pollen and nectar alone, for the flower-food available would be wholly inadequate to their requirements; but many species are occasional visitors to flowers.

Another great family of beetles are the Scarabæids, which

take their name from the genus *Scarabœus*, famous in art and Egyptian mythology. They are partly scavengers and partly leaf-chafers, comparatively few visiting flowers. Armies of rose-chafers (*Macrodactylus subspinosus*) often strip rose-bushes and other shrubs of both flowers and leaves (Fig. 90, No. 9), or devour the blossoms and ruin the crop of grapes. The common June-bugs defoliate trees, and on a warm evening the noise of their wings may be heard for a long distance. In these war-like days it is of interest to recall that a host of June-bugs once put British soldiers to flight near Boston. In John Trumbull's epic poem "M'Fingal" it is stated that, absurd as it may seem, it was a fact that some British officers, soon after Gage's arrival in Boston, while walking on Beacon Hill, shortly after sunset were greatly frightened by the sound made by flying June-bugs, which they took to be the sound of bullets. They left the hill in great haste, alarmed their camp, and later wrote home to England terrible accounts of being shot at with air-guns.

> "No more each British Colonel runs
> From whizzing beetles as air-guns;
> Think horn-bugs bullets, or through fears
> Musketoes takes for musketeers."

The snout-beetles or weevils (*Rhyncophora*), an immense group highly injurious to vegetation, seldom visit flowers, and as pollinators are of little importance. The long snout is used in excavating little pits in which they lay their eggs.

The wood-borers (*Cerambycidœ*), on the contrary, rank first in importance as flower-visitors among the families of the Coleoptera. They are present in great abundance on densely clustered small flowers, such as New Jersey tea (*Ceanothus*), viburnum, the cornels, spiræa, and chokeberry. They prefer nectar, which their mouth-parts, fringed with hair, enable them to lick up easily. As the grubs are tree-borers, the beetles

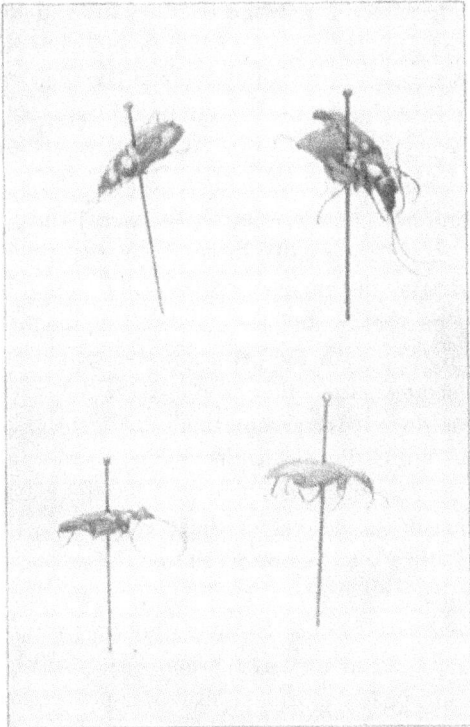

FIG. 92.   Beetles with a Tongue Resembling That of a Butterfly
Belonging to the Genus *Nemognatha*

Among the 100,000 or more described species of beetles, only two genera (*Nemognatha* and *Gnathium*) have a long sucking-tongue

are most common on flowers in or near woodlands, while on the same flowers a mile away they may be entirely absent. (Fig, 90, Nos. 1, 2, and 3.)   The elongated head and prothorax of many wood-boring beetles has been considered an adaptation for obtaining nectar; but, as the abdomen is also long and narrow, their cylindrical form has been determined more probably by their habit of gnawing burrows in solid wood, just as the elongated front of the head of the weevils has resulted from the excavating of little pits in which to lay their eggs.

Among the anthophilous, or flower-visiting Coleoptera, the two most remarkable genera are *Gnathium* and *Nemognatha*, of the blister-beetle family (*Meloidæ*), which have a slender suctorial tongue, like that of a butterfly, except that it cannot be coiled up. It varies greatly in length in the different species, attaining in one instance a length of 11 mm. Both genera live wholly on nectar, and they thrust this tongue in and out of tubular flowers with the precision and rapidity of bees. It may seem strange that other beetles have not acquired a suctorial tongue since it is common to all the butterflies and moths, but the Coleoptera did not begin visiting flowers till late in the history of their development, and they are dependent on nectar for food to such a small extent that variations in this direction would not be likely to be preserved in most cases. (Fig. 92.)

The primitive Coleoptera lived largely upon the ground, but as they learned to search for their prey on trees and herbage, they gradually began to visit flowers. They have never, however, been of much importance in flower-pollination, and floral structure has not been modified in any way as the result of their visits.*

* A complete list of the known anthophilous Coleoptera of New England with a description of the flowers visited by them will be found in two papers by the author, prepared with the co-operation of Mr. C. A. Frost, published in *Psyche; A Journal of Entomology*, vol. 22, No. 3, pp. 67–84; No. 4, pp. 109–117, 1915.

CHAPTER XII

POLLEN–FLOWERS

NOT long ago a popular youths' periodical published on its children's page a large picture of a climbing-rose bush from which a swarm of honey-bees was represented as gathering nectar. Beneath the bush was a still-house from which ran tubes to every flower. After the nectar had passed through a refining apparatus the bees were depicted as bottling and carting the honey away. It was an ingenious and amusing conceit, but unfortunately the roses do not yield nectar, and, alas! there is no such thing as rose-honey. The rose has proven a veritable thorn in the flesh to both artists and poets.

One of our popular poets sings of the honey-bee:

> "He harries the ports of the hollyhocks,
>   And levies on poor sweetbrier;
> And drinks the whitest wine of phlox,
>   And the rose is his desire."

Not at all. "He" (the worker-bee is an undeveloped female, and the drones do not visit flowers) does nothing of the kind, for the rose is nectarless and the phlox is a butterfly-flower. Before describing flowers the poet would do well to study them more closely.

Even bee-keepers, who should know better, very generally believe that bees gather nectar from the wild roses. "There has been some discussion of late," writes one of them, "as to whether bees get any honey from roses. I believe that I have seen them working very freely on wild roses, and I see no good reason why roses should not yield honey, as they belong to the same family as the apple, pear, plum, cherry, and raspberry. If one species in a family yields nectar we may expect that they will all do so." This may seem probable, but it is not the fact. In the buttercup family the buttercups, columbines, and larkspurs all secrete nectar, but the anemone and hepatica do not. Most species in the figwort family (*Scrophulariaceæ*) yield nectar, but some mulleins do not. In the honeysuckle family (*Caprifoliaceæ*) the honeysuckles and viburnum are nectariferous, but the elderberries are pollen-flowers. Some orchids secrete nectar, others do not. In the nightshade family (*Solanaceæ*) the nightshade is nectarless, but the ground-cherry (*Physalis*) yields nectar.

Although the handsome flowers of the rose are devoid of nectar, they contain such an abundance of pollen that they still attract a great many visitors, as honey-bees, bumblebees, leaf-cutting bees, mason-bees, ground-bees, as well as flies and beetles. Three or four little coal-black bees of the genus *Prosopis*, which look like ants, may be seen on a single rose eating pollen; but they are so small that they are of little use in pollination. But large bees, like the bumblebees, can hardly fail to come in contact with the stigmas, and thus are the most efficient pollinators. (Fig. 93.)

Conspicuous flowers pollinated by insects, which do not secrete nectar, are called pollen-flowers. Common pollen-flowers are the *Hepatica*, many species of *Clematis* and *Anemone*, the bloodroot, California poppy, the elders, rock-roses, loose-strifes, St.-John's-worts, poppies, nightshades, and species of mullein, *Spiræa* and *Thalictrum*.

While a part of the species of mullein (*Verbascum*) secrete a little nectar, others are nearly or wholly nectarless. The stamens are clothed with violet-colored hairs, which afford a good foothold to small bees and hover-flies while they are

Fig. 93. Wild Rose. *Rosa humilis*
A pollen-flower

gathering or eating the pollen. The mulleins appear to be in a transition stage. They have nearly ceased to secrete nectar, and have become dependent on their supply of pollen to attract insects, thus strongly suggesting that all pollen-flowers earlier in their history yielded nectar. (Fig. 94.)

The scarlet hue of the poppy has been said to repel bees, but bee-keepers who have seen their bees freely visiting these gaudy flowers do not need to have this assertion refuted. (Figs. 95 and 96.) Although an acre of poppies would not produce an ounce of honey, there are occasional reports of bees being stupefied by gathering nectar from poppy blossoms and lying about on the ground unable to fly; but all such stories are mythical. The hundreds of large, beautiful, purple flowers displayed by the garden-clematis (*C. Jackmanni*) contain no nectar, and bees visit them only occasionally to gather the scanty supply of pollen. But the wild clematis (*C. virginiana*) produces a profusion of small white flowers which are nectariferous. The yellowish-green, dilated filaments act as nectaries and secrete small drops of nectar on their inner surface. With age the filaments lengthen and turn white and then cease to produce nectar. Thus it is only the young blossoms which offer a sweet booty to their guests. From the bright-yellow flowers of the common loosestrife (*Lysimachia vulgaris*) a little, black, solitary bee (*Macropis ciliata*) gathers its entire supply of pollen for brood-rearing. (Fig. 97.)

The poppy and the rose produce a bountiful store of pollen, but otherwise remain passive, making no effort to place it on the visiting insect. On the other hand, there are many highly specialized pollen-flowers which possess varied devices for bringing the pollen in contact with the visitor. In the purple nightshade (*Solanum Dulcamara*), one of the simpler forms, the anthers unite in a cone around the style and a shower of pollen falls from pores in their tips, when a bee inserts its tongue between them. In the pulse family (*Leguminosæ*) there are a number of species, which although they have lost the power of secreting nectar still find the pumping and explosive mecha-

FIG. 94. Mullein. *Verbascum Thapsus*
A pollen-flower

FIG. 95. Red Poppy. *Papaver Rhœas*
A pollen-flower

nisms, previously acquired, very useful. In the flowers of the tick-trefoil (*Desmodium*), which are pollinated by bumblebees, the stamens are held under tension and shed their pollen while enclosed in the keel. When a bumblebee alights on the wings, the anthers are released and the elastic filaments project the pollen up in the air, as though there had been a slight explosion.

Fig. 96. Purple Poppy. *Papaver somniferum*
A pollen-flower

There is provision for only a single visit, since the entire stock of pollen is exhausted at once. Male bumblebees, which have no occasion to collect pollen, do not visit these flowers. The flowers of the lupine (*Lupinus*), another genus of the pulse family, are also nectarless. Here the pollen is expelled by a piston mechanism. The five outer stamens, after discharging their pollen inside the keel, wither up; while the five inner stamens act as a piston and push out the pollen during an insect visit. The blue lupine (*Lupinus subcarnosus*), the State flower of Texas, which carpets large areas of land with its handsome blue flowers, is visited by a great company of honeybees for pollen.

In some pollen-flowers there is a curious division of labor among the stamens, as in *Cassia* and *Heeria*. A part of the anthers, called nutritive anthers, are designed to furnish food to the visitors and are a conspicuous bright yellow; while a part reserved for pollination are an inconspicuous green or the color of the petals. In the mud-plantain (*Heteranthera reniformis*) there are two short stamens with golden-yellow anthers, and one long stamen with a pale-blue or greenish anther. While the bees are working on the short nutritive anthers, pollen is deposited on their bodies by the long reproductive anthers.

Pollen-flowers display every shade of color, as white, yellow, orange, red, scarlet, pink, purple, and blue. The differently colored varieties of the *Hepatica*, which may be found blooming in May, amid the brown leaves fallen from the trees during the season previous, are well described by Burroughs:

> "Sometimes she stands in white array,
>   Sometimes as pink as dawning day,
> Or every shade of azure made,
>   And oft with breath as sweet as May."

This variety in coloration is good evidence that the pollen-flowers were once nectariferous. There can be little doubt that at some time in their past history they all yielded nectar, and that subsequently this function was lost. The occurrence of isolated genera of pollen-flowers, as the roses in the rose family, the elderberries in the honeysuckle family, and the tick-

200

FIG. 97.   Common Loosestrife.   *Steironema ciliatum*
A pollen-flower

trefoils and lupines in the pulse family, in families where the majority of the genera yield nectar, admits of hardly any other explanation.   There is little difficulty in understanding how a flower may lose the power of secreting nectar.   Flowers from which the bee-keeper in one region derives an enormous surplus of honey may be in another wholly valueless. In the Rocky Mountain highlands alfalfa is the main reliance of the apiarist and no other honey-plant can compare with it, but east of the Mississippi River it is in most localities totally devoid of nectar. In the prairie States white clover is easily the foremost honey-plant, but in France one may ride for miles and not see a bee on the flowers. The flowers of the vine are in many places nectarless, but in others are reported to be nectariferous. Thousands of pounds of buckwheat-honey are annually produced in New York State, but farther west it is of little or no value. The flow of nectar in buckwheat is intermittent; in the morning it is active and the flowers are diligently visited by bees, but in the afternoon it ceases entirely and the bees remain idle in the hives. During the latter part of the day there is a great sea of white, fragrant flowers with hardly a bee on them. The secretion of nectar is evidently greatly influenced by soil and climate, and probably other factors, so that the loss of this function in certain plant genera is not at all surprising.

Wind-pollinated flowers, although they do not produce nectar, cannot be regarded strictly as pollen-flowers, since they rely on the wind for pollination. In early spring, honey-bees, hard-pressed to obtain pollen for brood-rearing, often, however, by thousands gather pollen from the alders, hazel-nuts, elms, hickories, and walnuts. Later in the season they may resort to the spindles of Indian corn and the green flowers of the ragweeds. Flies and beetles also feed on the pollen of rushes, grasses, and sedges.

CHAPTER XIII

IS CONSPICUOUSNESS AN ADVANTAGE TO FLOWERS?

FLOWERS usually owe their conspicuousness to a bright-colored corolla, as the rose and the buttercup. In the absence of petals the calyx may become bright colored, as in the clematis, anemone, marsh-marigold, and buckwheat; or both calyx and corolla may be highly colored, as in the columbines, larkspurs, and fuchsia. The catkins of the willows are rendered very conspicuous in early spring by the numerous yellow and red anthers, while in the meadow-rue the white filaments are broad and petaloid. The small leaves or bracts surrounding the flowers are also frequently brilliantly colored. In the painted-cup (*Castilleja*) the bracts are bright scarlet; in *Monarda media* they are purple, and in the bunchberry white (Fig. 98), while in the *Proteaceæ* of Australia the upper foliage leaves are blue.

Again, conspicuousness may be secured by massing small flowers in large clusters (Fig. 99), or by their production in great profusion. A single bluet is visible at a distance of only a few feet; but when they whiten a whole hillside they form a part of the facies of the landscape. When the dandelions bloom, whole fields become a bright golden-yellow in some New England towns; while in New Jersey large districts are white with daisy-blossoms, but unfortunately not for the harvest. On the prairies of Nebraska the ground-plum presents in spring a very striking appearance, the plants forming dense masses of reddish-blue flowers. In North Carolina, *Rhododendron*

Fig. 98. Bunchberry. *Cornus canadensis*
The small central flowers are rendered conspicuous by an involucre of four white leaves

maximum and *Kalmia latifolia,* or mountain-laurel, the two handsomest North American shrubs, "are seen to cover tracts of great extent, at one season presenting an unbroken landscape of gorgeous flowers." They adorn the valleys all around, says Asa Gray, in one of his letters, in immense abundance and

204

Fig. 99. Sunflower. *Helianthus annuus*
Conspicuousness is gained by the large size of the head, or capitulum

profuse blossoming, of every hue from deep rose to white. Almost equally conspicuous in various parts of the country are large areas brightly colored with yellow buttercups, golden-rods, sunflowers, orange hawkweeds, purple thistles, and blue lupines. In Texas the State flower, the blue lupine, carpets the ground for miles with innumerable blue flowers. But nothing in this world can surpass in beauty or lavish abundance the cloud-like masses of bloom displayed by the great northern apple-orchards.

"England has her furze-clad commons," says Wallace, "her glades of wild hyacinths, her heathery mountainsides, her fields of poppies, her meadows of buttercups and orchises—carpets of yellow, purple, azure-blue, and fiery crimson, which the tropics rarely can exhibit. We have smaller masses of color in our hawthorn and crab-trees, our holly and mountain-ash, our broom, foxgloves, primroses, and purple vetches, which clothe with gay colors the whole length and breadth of our land. They are characteristic of the country and climate, they have not to be sought, for they gladden the eye at every step."

Brilliantly colored flowers usually contrast with the green foliage of trees, or of herbaceous plants, or with the grass. But the white and blue hepaticas, which bloom with the opening of the new season, have for a background the sere and brown leaves, fallen from the trees during the preceding autumn; and contrasting with the dark soil in dense woods gleams the snow-white Indian-pipe. (Fig. 100.) Flowers which rest upon the surface of the water are often white or yellow, as the yellow and white water-lilies. Nocturnal flowers are also generally white or yellow, since purple or blue would be invisible in the darkness of night.

In Europe and North America, and in all lands where there

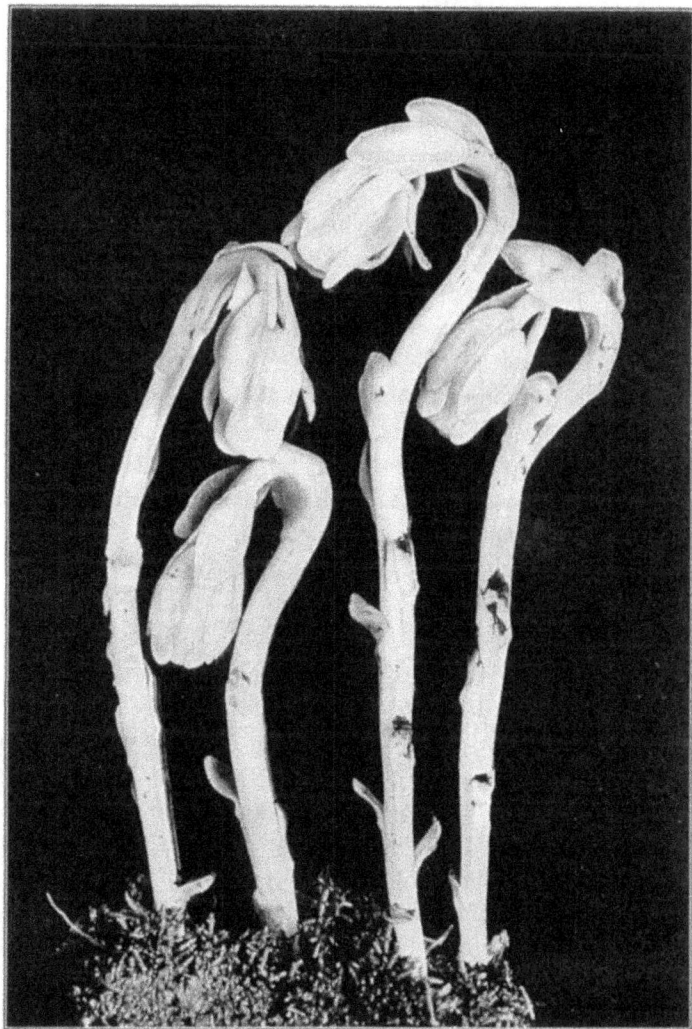

FIG. 100. Indian-Pipe. *Monotropa uniflora*

The snow-white parasitic plants contrast strongly with the dark soil of the woods in which they grow

is an insect fauna rich both in species and individuals, flowers display an infinite number of brilliant hues and delicate shades which surpass the power of the artist and naturalist to describe. There is a wonderful variety of bicolored, tricolored, and variegated blossoms, often mottled and veined in endless ways. Not only are the prismatic colors—red, orange, yellow, green, blue, and violet—displayed by many species with a profusion of intermediate shades, but rarer colors like black, brown, scarlet, crimson, and lurid purple are not unrepresented. Nature has tried her skill as a colorist in the metallic lustres and translucent hues of minerals; in the vivid, living tints of corals and sea-anemones; in the lights and shades reflected by the scales of the butterflies' wings; and in the brilliant iridescent plumage of birds; but nowhere are her inexhaustible resources in chromatics so bountifully displayed as in the colors of flowers.

A flora in which the flowers were all of one color would be at a great disadvantage. The value of color contrasts is evident, for they enable the visitors, more especially the bees, easily to remain constant to a single plant species in collecting pollen and nectar. If they were to visit flowers indiscriminately, much pollen would be wasted and much time and effort lost in locating the nectar. In the Alpine flora of the Tyrol, in the heights above the tree-line, there is no spring and no autumn— only a short summer following a long winter. All the flowers have, therefore, to blossom in this short time. "White and red, yellow and blue, brown and green," says Kerner, "stand in varied combination on a hand's breadth of space. Hardly has the snow melted than there appear almost simultaneously the violet bells of the soldanellas and the golden flowers of the cinquefoil, the white crowfoot and androsace, the red silenes and primulas, the blue gentians and the yellow auriculas, the heaven-blue forget-me-not and the yellow violet, as well as the saxi-frages, in every conceivable color." Such a meadow in Alaska, where the summers are equally short, is well described by Burroughs:

> "Starred with flowers of every hue,
> Gold and purple, white and blue;
> Painted cup, anemone,
> Jacob's ladder, fleur-de-lis,
> Orchid, harebell, shooting-star,
> Crane's-bill, lupine, seen afar;
> Primrose, poppy, saxifrage,
> Pictured type on Nature's page."

According to a well-known principle of physics, each color appears more brilliant in contrast with other hues than it would if viewed alone. This can be easily shown by a simple experiment, which any one can perform. Cut out two pieces of red paper, each two inches square. Place one of the red squares on a large sheet of green paper and the other red square on a large sheet of red paper. The red square on the green paper will appear so much more brilliant than the red square on the red paper that the observer will have difficulty in believing that they are identical in hue. (Figs. 101 and 102.)

Is this beautiful and varied display of coloration by flowers of no use? Has it no more significance than the vivid iri-descent hues of minerals and precious stones? Is it merely an incidental result? To most observers it has long seemed self-evident that conspicuousness is a manifest advantage. If insects possess a well-developed sense of vision, bright colors cannot fail to be of benefit to them as well as to flowers by en-abling them easily to discover isolated blossoms and to econ-omize time by being faithful to single plant species. No argu-ment is needed to prove that such a correlation is desirable, and that, if it is non-existent, its absence registers a failure on Nature's part to make the most of an opportunity.

According to the teachings of Sprengel, Darwin, and Mueller,

FIG. 101. Bean. *Vicia Faba*
Color contrast, a black spot on the wings of the papilionaceous
white corolla

and most other flower-biologists, the bright hues of flowers serve as signals, or flags, to attract the attention of insects living on pollen and nectar. The more conspicuous a flower, or flower-cluster, the better are its chances of pollination.

FIG. 102. Purple Coneflower. *Echinacea angustifolia, a*
Coreopsis. *Coreopsis tinctoria, b*

Color contrast, the disk flowers and the lower part of the rays are brown-purple, the upper
part of the rays yellow

Many colors are better than one, since the flowers are rendered more conspicuous by contrasts with each other as well as with foliage, and insects are less liable to visit them indiscriminately. The various floral colors have been evolved by the selective agency of insects, especially bees, which are able easily to distinguish between them, and in the absence of visitors flowers would have remained green, or dull-colored, similar to wind-pollinated blossoms. In some instances Mueller believed that the visitors manifested a preference for certain colors, as honey-bees for blue, butterflies and humming-birds for red, hover-flies for yellow, and carrion-flies for lurid purple; but in the light of more recent investigations it may be doubted whether insects receive more pleasure from one color than another. The usefulness of floral-color contrasts is sufficient to explain their development without recourse to the supposition that they afford an æsthetic pleasure to insects.

With the exceptions of the criticisms of Bonnier, in 1879, Mueller's doctrine remained almost universally unquestioned until 1895, when Felix Plateau, of the University of Ghent, made the sensational assertion that Mueller had been misled by a too vivid imagination, and that in the mutual relations of insects and flowers the bright colors of the floral leaves have not the important rôle that he had attributed to them. All the flowers in nature might be as green as their leaves, without their pollination being compromised. It is not their sense of color but their sense of smell which enables insects to discover flowers which contain nectar and pollen.

Assertions so revolutionary were naturally received with much incredulity, and in some instances, as Plateau naïvely remarks, were criticised with merciless severity. While replying to his critics with admirable courtesy, Plateau constantly sought for new evidence, and actively maintained his views

to the close of a long life. Had he been content to refute
Mueller's theory that insects exhibit color preferences, it is
probable that he would have met with little opposition; but
his sweeping denial of the value of conspicuousness in any
degree to flowers has not met with general acceptance and
can be easily shown to be incorrect.

If the flowers of the common pear (*Pyrus communis*) be deprived of their petals, honey-bees will at once cease to visit
them for nectar, as is shown by the following observations. A
cluster of seven blossoms near the end of a branch was watched
for fifteen minutes and received eight visits from honey-bees.
The petals were now all removed and it was observed for a
second quarter of an hour. Though a number of bees flew
near by, it received not a single visit. During a third fifteen
minutes there were two visits, due in part to association, for
the bees came from other blossoms on the same tree, which had
proved the first source of attraction.

Two other clusters of flowers, growing side by side, but nearer
the bole of the tree, consisting each of 8 flowers, were observed
for fifteen minutes, and 16 visits of honey-bees were noted. The
petals of one of these clusters were now removed. During
fifteen minutes the adjacent cluster, which still retained its
petals, received 11 visits, while not one was made to the cluster
without petals. In one instance a bee hovered over it but did
not alight. These results were very conclusive, and showed
that the bees were guided almost entirely by the presence of
the petals.

Similar results were obtained from an experiment with two
groups of flowers belonging to the common borage (*Borago
officinalis*, Fig. 103.) They were distant apart about 6
inches; one contained 5 flowers; the other, which was at a little
higher elevation, contained 4 flowers. They were both watched

for ten minutes. The first received 15 visits from honey-bees, the second 13 visits. The blue corollas, together with the cone of black anthers, were now removed from the flowers of the first group. The two groups were now observed for a second ten minutes; the first received no visits, the second 7 visits from honey-bees. Once a bee hovered around the denuded flowers of the first group, but failed to alight, although they contained an abundance of nectar. There were scattered upon the ground many partially withered corollas and it was interesting to notice that a bee was twice seen to fly down toward them. The value of conspicuousness was here again very clearly established.

A staminate flower of the garden-squash (*Cucurbita maxima*) was placed under observation for ten minutes and received 12 visits, 8 from honey-bees and 4 from bumblebees (*Bombus terricola*). The yellow corolla was then removed, and it was watched for a second ten minutes, during which it received only a single visit from a bumblebee. Two squash-flowers, both staminate, growing side by side, their corollas touching, were then selected. Both were observed for ten minutes. Number one received 6 visits—4 from bumblebees, 2 from honey-bees; while number two received 13 visits, all from bumblebees. The fresher condition of the second flower probably accounted for the larger number of visits. The yellow corolla was now cut away from number two, and both flowers were watched for another ten minutes. No visits were made to the denuded flower, but number one received 12 visits, 6 from honey-bees, and 6 from bumblebees. In the previous experiments the number of visits to the complete flowers were numerous and decisive. On the contrary, they ceased almost entirely to the decorollated flowers, although they contained an ample supply of nectar. That the white, blue, and yellow corollas were bene-

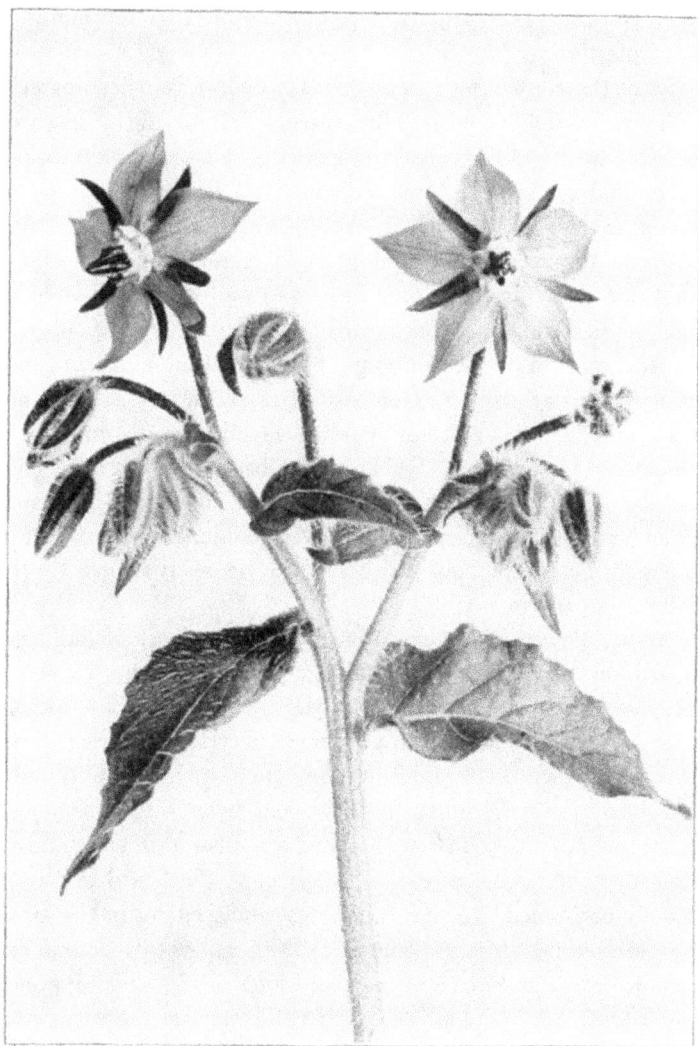

Fig. 103. Borage. *Borago officinalis*

When the petals were removed honey-bees at once ceased to visit the flowers

ficial to the flowers of their respective species does not admit of any question.

Since there are a few green flowers, which secrete nectar freely and are frequently visited by insects despite the absence of bright colors, as the garden-asparagus, basswood and woodbine, Plateau argued that, therefore, all flowers might be as green as their leaves without diminishing the number of insect visits. But a careful examination of greenish flowers shows that for the most part they are small and wind-pollinated or self-fertilized and are never, or only rarely, visited by insects. In the case of the exceptions, which contain an abundance of nectar, they will be often visited after the nectar has once been found, but it will not be found as quickly as it would be if they were conspicuous. When honey-bees are given the choice between a conspicuous and an inconspicuous object, both supplied with honey, they will discover first and visit more frequently the conspicuous object, as can be easily shown by the following experiment:

About 25 bees were accustomed to visit a piece of dull-gray board on which a small quantity of sugar-syrup had been placed. Three feet away from the board there was laid on the grass of the lawn a dried yellow everlasting-flower (*Helichrysum bracteatum*) containing a small quantity of honey. On the opposite side of the board, 3 feet away, there was placed a green apple-leaf, on which there was also a small quantity of honey. As soon as the sugar-syrup of the board had been wholly consumed the bees began describing circles in the air in a search for a further supply. They repeatedly found the yellow flower and at one time there were 3 bees sucking honey on it; but not a single bee found the honey on the apple-leaf. According to the reiterated statement of Plateau all flowers might be as green as their leaves without their pollination being compromised, and color and form are of little consequence in comparison with odor. But this experiment and many others showed that color contrast was of great value—in this particular experiment it was indispensable. If the leaves provided with an ample supply of honey could not obtain a single visit, how little chance would there be for an isolated plant with small green flowers growing in a secluded location attracting visitors! But a bright-colored flower in the same locality would be much more likely to gain the attention of pollinators.

From the preceding experiments it appears that as soon as a conspicuous flower loses its petals bees cease to visit it, and that they find a bright-colored flower more quickly than they do a green one. It remains to show that they can distinguish between different colors, for, if they cannot, then, a polychromatic flora possesses no advantage over one in which the flowers are all of the same hue. For this purpose we have made use of differently colored flowers of the same species, which are alike in form and odor, and differ only in color. Differently colored slips of paper might also be employed.

A honey-bee was trained to visit a purple sweet pea, on the wings of which honey had been placed. The flower was laid on a dull-colored board, raised several feet above the ground. After the bee had become accustomed to the purple color, while it was absent at the hive, the purple flower was moved 12 inches to the right and a red sweet pea with honey on the wings was put in its place. The bee returned to the purple flower and after taking up a load of honey left again for the hive.

During its absence no change was made in the position of the flowers. The bee on its return hovered over the red sweet pea and alighted on it for a moment or two, but then left for the purple flower where it took up its load of honey.

THE FLOWER AND THE BEE

While the bee was away the flowers were transposed, the red blossom being put in the place of the purple, and the purple in the place of the red. The bee returned to the purple flower.

After the bee had left for the hive the flowers were again transposed. On its return the bee manifested a little hesitation, but soon went to the purple blossom.

While the bee was absent, the flowers were still again transposed, but on its return it flew directly to the purple flower.

Although the experiment was continued further, it is not necessary to give additional details, since it is clear that the honey-bee was able to distinguish the purple sweet pea from the red one. In the same way a honey-bee showed that it was able to distinguish between the red and yellow flowers of *Portulaca grandiflora;* and the greenish white and purple flowers of *Cobœa scandens.*

In another experiment blue and red slips of paper 3 inches long by 1 inch wide were used instead of flowers. After the bee had made a few visits to the blue paper, on which there was a small quantity of honey, the red slip of paper with a little honey on it was placed 6 inches to the right of it. The bee returned to the blue paper, which still remained in its original position. The blue and red papers were now transposed 9 times and in each instance the bee returned to the blue paper, from which it gathered its load of honey. In another experiment a honey-bee distinguished with equal ease between blue and yellow slips of paper. Bee-keepers have long recognized the ability of bees to distinguish between different colors, and at times paint their hives red, white, and blue in order to prevent young queens from entering the wrong hive after mating.

If bees can so easily distinguish between different colors, how, then, does it happen that they so often visit indiscriminately in our gardens the differently colored varieties of the same species of flower, as the white, yellow, orange, red, and purple varieties of *Zinnia*, or the red, white, blue, and purple varieties of bachelor's-button (*Centaurea cyanus*)? It is obvious that the flowers belonging to each species are alike in shape, odor, and nectar, and differ in hue alone. Under these circumstances it is for the advantage of bees to pass freely from one color to another, and this they speedily learn to do.

Since bees are able to distinguish between different colors, and cease to visit flowers as soon as the brightly colored floral leaves are removed, and find a flower which contrasts sharply in color with green foliage more quickly than one which is similarly colored, conspicuousness is clearly a great advantage in attracting insects.

If, however, brightly colored flowers, as in the case of many gaudy exotics of cultivation, are nectarless and yield little or no pollen, bees soon learn that no food is to be obtained from such blossoms, and remembering this, thereafter visit them only occasionally. The large flowers of the cultivated purple clematis (*Clematis Jackmanni*), for example, are nectarless and odorless, but produce a small amount of pollen. Careful and almost continuous observation showed that they were at times visited by honey-bees and solitary bees, which gathered all of the pollen. I inspected the flowers many times without finding any insects, and a casual observer might easily conclude that they were entirely neglected. After the pollen had been removed there was no reason why insects should continue their visits.

For the purpose of learning whether the visits of bees could not be induced in large numbers, I next placed on a few of the flowers sugar-syrup, which is an odorless sweet liquid. Honey-bees soon began to visit the flowers, and continued to come in increasing numbers so long as I supplied the syrup. This experiment was repeated with many other garden-flowers which were nectarless with similar results. Wild and field flowers also which in one locality freely secrete nectar, as alfalfa, white clover, buckwheat, and goldenrod, and are visited by many insects, are sometimes in other localities nectarless and almost entirely neglected. Insects, therefore, perceive the colors and forms of neglected flowers, and the rarity of their visits is the result of their memory of the absence of food materials.*

* Readers desiring to pursue this subject further are referred to the following articles by the author: "Is Conspicuousness an Advantage to Flowers?" *Amer. Nat.*, vol. 43, pp. 338–349, 1909. "Can Bees Distinguish Colors?" *Amer. Nat.*, vol. 44, pp. 673–692, 1910. "The Pollination of Green Flowers," *Amer. Nat.*, vol. 46, pp. 83–107, 1912. "Conspicuous Flowers Rarely Visited by Insects," *Jour. Animal Behavior*, vol. 4, pp. 147–175, 1914. "The Evolution of Flowers," *Scientific Monthly*, vol. 4, pp. 110–119, 1917.

CHAPTER XIV

THE COLORS OF NORTH AMERICAN FLOWERS

THE distribution of coloration in our flora is a question of much interest, but one which up to the present time had received very little attention. Some years ago I began an inquiry as to how many flowers there are of each color in the flora of North America. In northeastern America, north of Tennessee and east of the Rocky Mountains, there have been described 4,020 species of flowering plants, or Angiosperms. Partly by direct examination and partly by reference to various systematic works, I have tabulated the entire number according to the predominant colors of their flowers. I find that in the area named there are 1,244 green, 956 white, 801 yellow, 260 red, 434 purple, and 325 blue flowers. In every hundred species there are 30.9 green, 23.8 white, 19.9 yellow, 6.4 red, 10.9 purple, and 8 blue flowers. Their distribution among the different flower series is shown in the following table:

Series	Green	White	Yellow	Red	Purple	Blue	Total
Monocotyledons.....	857	82	41	22	22	34	1,058
Dicotyledons—							
Choripetalæ:							
Apetalæ........	175	89	51	45	24	....	384
Polypetalæ......	140	410	333	84	193	57	1,217
Gamopetalæ......	72	375	376	106	198	234	1,361
Total......	1,244	956	801	257	437	325	4,020

The green, white, and yellow flowers number 3,001, or three-fourths of the entire number; while the red, purple, and blue amount to only 1,019. Although there are many exceptions, especially in the pulse, mint, and figwort families, the first group contains largely regular, rotate, or tubular flowers with the nectar accessible to a large miscellaneous company of insects, as beetles, flies, butterflies, wasps, and bees. Yellow irregular or bilabiate flowers seem to be often the result of the greater persistence of the primitive yellow pigment, and its little tendency to vary with the specialization of the corolla. Many white irregular flowers are undoubtedly due to reversion. The flowers belonging to the second group are very frequently irregular or bilaterally symmetrical, with the nectar concealed, and are chiefly attractive to long-tongued bees, butterflies, and flies. The tendency of flowers to change from green, white, and yellow to red, purple, and blue, is much stronger than the reverse; but red, purple, and blue flowers usually have the petals white or yellowish at the base and in the bud, and not infrequently the whole corolla reverts to one of these colors.

Have these relations any significance? Undoubtedly they have. They are signals pointing out to us the course our flora has pursued in its evolution. The green, white, and yellow colors are older and more primitive than the red, purple, and blue, and were much more common in the primordial flora. The red, purple, and blue flowers are, as a whole, of much more recent origin, and have been developed from green, white, and yellow blossoms. For example, the buttercups are a much older genus than the columbines or larkspurs, and the cinquefoils are more ancient than the pea, bean, or vetch; while again the viburnums are older than the honeysuckles. The orchids have certainly developed more recently than the lilies. Occasionally irregular flowers revert to their ancestral stages and produce perfectly regular forms. These color changes are often recapitulated by individual flowers; white corollas changing to red, as in the sweet-william, or to yellow, as in the climbing honeysuckle, or from yellow to red, as in lantana and the flowering currant (*Ribes aureum*), or from red to blue, as in the forget-me-not (*Myosotis versicolor*).

Let us next inquire how many of these 4,020 flowers found in northeastern America are pollinated by the wind and how many by insects. Among the wind-pollinated plants are the grasses, sedges, and rushes; many homely weeds like the pigweeds, sorrels, nettles, and ragweeds, as well as many deciduous-leaved bushes and trees, as the alders, poplars, elms, beeches, and birches. After a careful examination of every genus I place the number of wind-pollinated plants (including a few pollinated by water) at about 1,046. This number is, perhaps, a little too large, for in the case of some Western species there are no recorded observations and they may be self-pollinated. Still it cannot be far from correct, since the grasses and sedges alone in this area include 705 species, the rushes 47, the pondweeds (including 8 water-flowers) 42, the deciduous-leaved trees and shrubs 71, the chenopods and amaranths 54, and 36 species in the *Compositæ*.

Wind-pollinated plants have usually small and inconspicuous flowers which are green or dull-colored, and which flower and fruit entirely unnoticed. It would be of no advantage to them to produce bright colors or sweet odors, for the wind bloweth where it listeth regardless of all such attractions. The birches, however, have golden and greenish-yellow aments, and the blossoms of the elm are purplish. The glumes of grasses and the perianths of rushes are also often purplish or reddish. So conspicuous are the flowers of some rushes that they attract a few insects. The sorrels may have the entire plant red-colored, and butterflies may seek nectar in the flowers. The plantains are midway between wind-pollination and insect-pollination, and some species display several hues and are pleasantly scented. But as a whole wind-pollinated plants have small, greenish flowers.

Setting aside the great company of dull-colored, wind-pollinated flowers, there remain in northeastern America 2,972 species which are pollinated by insects or are self-pollinated. Of this number 223 have green, 955 white, 790 yellow, 257 red, 422 purple, and 325 blue flowers.

GREEN FLOWERS

The primitive color of flowers was doubtless green. If the theory of the poet Goethe that the flower is a metamorphosed bud, or part of a branch of leaves, be admitted, this is self-evident. Despite many attacks, this doctrine has never been disproven, at least historically. In most flowers the calyx has remained green, and in some genera, as *Hepatica*, its derivation from leaves is evident from inspection. It is not uncommon in the buttercups, anemones, poppies, mustards, tulips, and many other genera for both the sepals and petals to revert to green leaves, and I have before me a flower of *Fuchsia* with three white petals, while the fourth is a green leaf. In *Cactus* no line of demarcation can be drawn between bracts, sepals, and petals, and all three are in the same spiral series. Even assuming that foliage-leaves were derived from sterile spore-bearing organs (sporophylls), there is every reason to believe that the sheathing bracts of the earliest flowers were green. In the Black Hills a fossil "flower" of a cycad-like plant (*Cycadeoidea*) has been found by Wieland, which is protected by an indefinite number of hairy, green, bract-like leaves. (Fig. 104.)

The green hue of both green leaves and flowers is produced

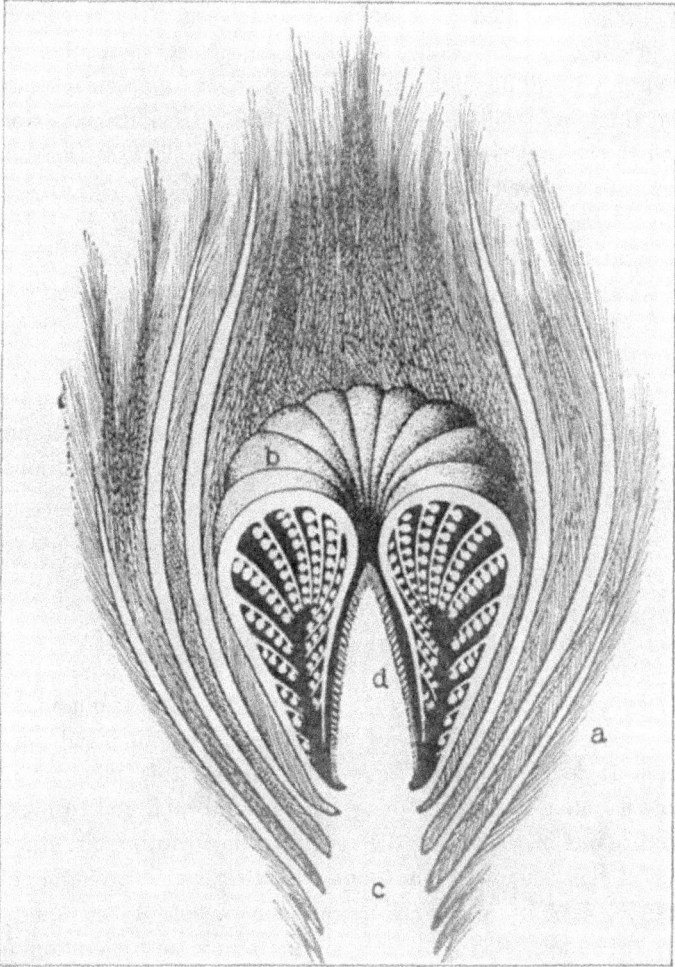

FIG. 104. Diagram of the "Flower," or Strobilus, of *Cycadeoidea dacotensis,*
a Fossil Plant from the Black Hills, South Dakota

a, Hairy, green sheathing bracts; *b,* folded stamens; *c,* elongated axis; *d,* conical mass of
sterile and fertile scales, the latter bearing terminal naked seeds. From somewhat
similar ancestors modern flowers were perhaps derived. (After Wieland)

by a pigment called chlorophyll, or leaf-green. If a few leaves
of grass, or of any common plant, be placed in alcohol the
chlorophyll will dissolve out, forming a yellowish-green solution,
and the leaves will be left entirely white. Chemical examination shows that there are two kinds of chlorophyll in the solution, a blue-green, which is abundant, and a yellow-green pigment which is less common. Place this solution in bright
sunlight and the green color will soon be destroyed. Green
seaweeds, when left on the beach by the waves, soon turn
yellowish owing to the destruction of the chlorophyll. In living leaves and green flowers under the action of bright light the
green pigment is constantly being destroyed and renewed, so
that no two leaves are identical in hue, and no leaf long remains
the same shade. Four hundred years ago a German poet,
Freidank, observed this fact.

> "Many hundred flowers
> Alike none ever grew;
> Mark it well, no leaf of green
> Is just another's hue."

Leaf-green, or chlorophyll, is not only the most common, but
it is also the most useful of all pigments, for all life depends
upon it for existence. Leaves containing this pigment are
able to make use of the energy of the sunbeam, and to manu-
facture out of water and the carbonic dioxid in the air, starch,
one of the principal plant-foods. That is, out of mineral substances they build up an organic substance. As all animals
are dependent either directly or indirectly on vegetation for
support, the destruction of chlorophyll would mean the disappearance of life from the earth. All living beings are dependent upon chlorophyll and the radiant energy of the sun.
"In this sense," says Tyndall, "we are all souls of fire and

children of the sun." In the making of starch the leaf-factories absorb most of the red, orange, and blue rays of light, but make no use of the green, and hence the color of foliage is green, a most fortunate result, since a landscape of any other hue would have been almost intolerable.

Most of the 223 insect-pollinated or self-pollinated green flowers in northeastern America are small or even minute, as in the pinweeds (*Lechea*). Many have no petals and their color is due to the green calyx, as 15 species of the buckwheat family, 8 species of sandworts and chickweeds, several species of the rose family, the rock-maple (Fig. 105), and the water-purslane. Many green flowers show undoubted evidences of retrogression, as numerous spurges, the water-milfoils where the sepals and often the petals have been lost; while the composite flowers of the wormwoods (*Artemisia*) and ragweeds (*Ambrosia*) have retrograded until they have become wind-pollinated.

The flowers of the vine family depend chiefly upon their fragrance to attract insects. The green petals never expand but fall away by separating at the base and coiling spirally upward. The fragrance which resembles that of mignonette can be perceived at a long distance. Kerner relates that in a journey up the Danube he found the whole valley of the Wachan so filled with the scent of vine-flowers that it seemed impossible that they could be far off, yet the nearest vines were 300 yards from the boat.

While, in general, green flowers are visited chiefly by flies, beetles, and the smaller bees, as *Clintonia borealis*, the large yellowish-green panicles of the false hellebore (*Veratrum viride*), and the smaller clusters of the smilax family, there are a few species which secrete nectar very abundantly and are visited by great numbers of insects. Basswood is one of the most valuable honey-plants in North America, and its greenish flowers yield annually thousands of pounds of a rich, aromatic honey. The rock-maple in early spring, and the Boston ivy and woodbine (Fig. 106) later in the season, are also valuable sources of nectar to the bee-keeper. Large green flowers occur in various exotic species of the nightshade family and in some Brazilian orchids. They are strongly scented in the evening and are attractive to nocturnal moths.

Green flowers often contain other pigments besides chlorophyll, or leaf-green, as carrotin, tinging them various shades of yellow, or green granules may be mingled with violet-colored sap as in the dull-purple corolla of belladonna (*Atropa belladonna*), while the brownish color of the gooseberry is due to red-cell sap and chlorophyll.

YELLOW FLOWERS

The green pigment, or chlorophyll, in leaves is invariably accompanied by two yellow pigments, carrotin, so-called because it is common in the root of the carrot, and xanthophyll, or leaf-yellow. Carrotin, to which most yellow flowers owe their hue, is a solid substance, occurring in petals in small round granules called plastids. It is very widely distributed in seaweeds, fungi, lichens, mosses, ferns, and the higher plants, in autumn leaves and in fruits and seeds. It is insoluble in water, but readily soluble in ether. The yellow plastids of flowers are not always round, but are sometimes angular as in the garden-nasturtium. In the tomato, asparagus, thorn-bush, and in some species of rose, the plastids of the fruit are spindle-formed, or irregular-shaped, and are fire-red, orange-red, or yellowish red. In yellow leaves the plastids are round; but in autumnal leaves they occur in irregular masses.

The scarlet poppy, tulip, and fire-red canna owe their colors

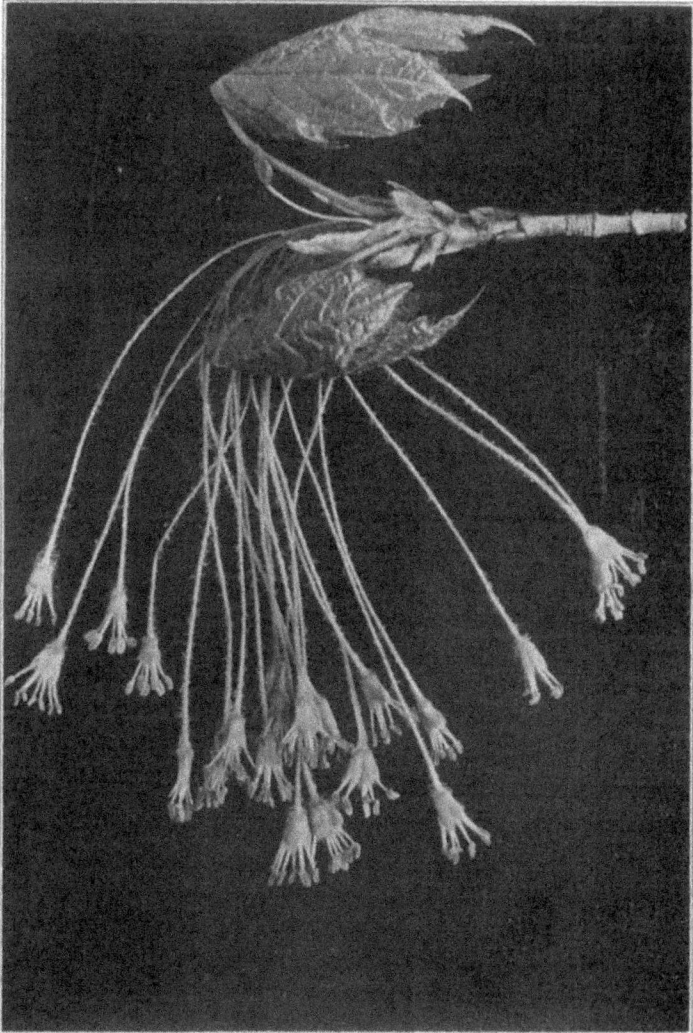

FIG. 105. Yellowish-green Flowers of Rock-Maple. *Acer saccharum*

to a mixture of yellow plastids and red cell-sap. On the other hand, dingy or dull colors result from a combination of violet sap with yellow granules. Carrotin is much less sensitive to the effects of light than chlorophyll, as may be readily shown by the following experiment: If the carrotin contained in a few slices of carrot-root be dissolved out in ether, the yellow solution will not lose its color under ten days, while the green hue of a solution of chlorophyll will disappear in twenty-four hours.

Yellow was doubtless one of the first colors displayed by the flowers of the primitive flora; and, in view of the wide distribution of yellow pigments in leaves, the development of yellow petals offers little difficulty, and many leaves, fruits, and flowers afford suggestions as to the way in which this change might take place. The quantity of yellow pigment in the foliage of different plants varies greatly; while in some species it is scarcely perceptible, in others it is so abundant as to tinge the whole plant yellow, and in a few golden-yellow species the chlorophyll appears to be nearly or wholly excluded. Many fruits change from green to yellow in ripening, and yellow coloring is especially prominent in foliage both in spring and autumn. Not infrequently the change from green to yellow occurs in flowers after they have opened, as in penny-cress (*Thlaspi*) and bitter cress (*Cardamine*); in the yellow water-lily (*Nymphœa advena*) the outer sepals are half green and half yellow; while yellow tulips in expanding display every shade from dark green to bright yellow. In double English buttercups in the bud the petals are wholly green, becoming later golden yellow. Every stage of the transition from green to yellow is constantly illustrated by fruits and flowers.

According to Stahl it is more costly and difficult to produce chlorophyll, which requires for its formation the presence of

230

nitrogen, phosphorus, potassium, magnesium, and iron, than carrotin, which is composed of only carbon and hydrogen. Anything, therefore, which retards or prevents the formation

FIG. 106. Woodbine. *Psedera quinquefolia*
Small green flowers

of chlorophyll, as the absence of the proper elements, will cause the flower to become yellowish; and, as has been shown, carrotin is much more persistent in intense sunlight than the green pigments. In the beginning of entomophily (adaptation to insects) there would naturally be from time to time more or less yellow flowers in the ancient flora, and the advantage afforded by greater conspicuousness would soon cause them to become common.

Among the flowers which owe their yellow color chiefly to carrotin are the *Abutilon*, *Adonis*, squash, *Forsythia*, sunflower, jewelweed (*Impatiens biflora*), *Kerria japonica*, evening-primrose, yellow roses, dandelion, and nasturtium. The yellow pigment xanthophyll does not occur in plastids, but is dissolved in the cell-sap. To it are due the yellow color of the peel of the lemon, the yellow flowers of the dahlia, butter-and-eggs (*Linaria vulgaris*), snapdragon, and all the yellow-flowering thistles as well as many other flowers. (Fig. 107.)

There are 790 yellow flowers in northeastern America, which vary in size from the large campanulate cups of the squash to the small flowers of the creeping buttercup. Usually they are wheel-shaped, as in the buttercups and fivefingers; but not infrequently they are very irregular in form, as in the pea and figwort families, where the corolla bears a more or less fancied resemblance to a butterfly or the head of a reptile. As a whole, however, they are much less specialized than red and blue flowers. Irregular yellow flowers probably owe their hue largely to the great persistency of the yellow pigment carrotin. Both yellow and white flowers are common in primitive families. For instance, in the buttercup family there are 38 yellow and 26 white flowers; in the mustard family, 46 yellow and 54 white; in the rose family, 39 yellow and 35 white.

While trees and shrubs with white flowers abound everywhere, trees and shrubs with yellow flowers are comparatively rare. A number of common trees have small yellowish or greenish-yellow flowers, as the rock-maple, striped maple, chestnut, and basswood; while among shrubs there are the barberry, fly-honeysuckle, jessamine, and bush-honeysuckle. Familiar

FIG. 107. Cucumber. *Cucumis sativus*
The yellow flowers owe their color to plastids of carotin in the cells

yellow-flowered shrubs under cultivation are the *Forsythia*, golden currant, and yellow rose. The yellow hue of the willows is due to their anthers, for they have no perianths. Most plants with yellow flowers are herbaceous. When the blossoms are of small size they are usually assembled, like small white flowers, in clusters, as in the mustard, saxifrage, carrot, and thistle families. (Fig. 108.)

In the pink family, although there are 56 white flowers, there are no indigenous yellow species; and in the aquatic water-plantain family the entire 19 species are white; but, on the other hand, in the St.-John's-wort family there are 22 yellow and 2 red flowers, while white fails entirely. Yellow is very common among the primroses and nightshades, but rare among the heaths and gentians. Irregular, or zygomorphic, yellow flowers are much rarer than regular forms, and are most common in the orchis, pea, violet, figwort, and honeysuckle families, which contain a total of 104 species.

In the aster or thistle family (*Compositæ*) there are 262 yellow flowers and 134 white. Though this is the highest of plant families, the central florets of each head are very small, and the corolla has been very little modified; consequently the primitive yellow has been largely retained. Some large genera, as golden aster (*Chrysopsis*) and goldenrod (*Solidago*) with one exception, and groundsel, or *Senecio*, have all the flowers of this hue. So abundant are many yellow-flowered species in various localities, as the sunflowers, goldenrods, Coreopsis, Spanish needles (*Bidens aristosa*), gum-plant (*Grindelia*), crownbeard (*Verbesina*), marigolds, and dandelions that yellow is more predominant in the floral landscape of North America than any other color. Yellow might well be our national color, and the goldenrod our national flower. It is the most bright and cheerful of colors since it reflects the largest amount of light, and it is

FIG. 108. Garden-Marigold. *Calendula officinalis*
The most familiar yellow flower of cultivation

doubtless for this reason that yellow flowers enjoy so great popularity both in the United States and Europe.

The goldenrods, a genus of beautiful and stately plants, which are everywhere common in North America, bloom from midsummer until late fall. They are most valuable as a source of honey, and in New England are the main reliance of the bee-keeper for winter stores for his colonies. They are great favorites with the honey-bee, and are visited also by more than 100 other species of insects. The bright-yellow color of the flowers renders them conspicuous both by day and evening; and as the temperature of the inflorescence at night is several degrees above that of the surrounding air, they sometimes serve as a nocturnal refuge for insects.

> "And in the evening, everywhere,
>   Along the roadside, up and down,
> I see the golden torches flare,
>   Like lighted street-lamps in the town.
>
> I think the butterfly and bee,
>   From distant meadows coming back,
> Are quite contented when they see
>   These lamps along the homeward track."
> —*Sherman.*

Yellow flowers in their natural state exhibit but little variation in color. They change most readily to white, and less often to red and blue. Under cultivation Darwin noted a double yellow hollyhock, which suddenly turned one year into a single white form, and a chrysanthemum has been observed to bear both yellow and white flowers. Some species of mustard regularly fade to white, and not a few white flowers show that they are derived from an ancestral yellow by retaining vestiges of this color at the base of the petals, as the water-crowfoot. The pale-yellow flowers of *Œnothera laciniata,* of the golden

236

currant (*Ribes aureum*), and of the bush-honeysuckle (*Diervilla trifida*) in fading change to rose or red. A species of forget-me-not (*Myosotis versicolor*) is at first yellow, changing later to sky-blue. In the violet family the smallest and simplest species is yellow, and the most highly specialized is blue, while all the intermediate stages are shown by the pansy.

WHITE FLOWERS

White flowers, of which there are 955 species in northeastern North America, are most common in our flora as well as in that of Europe. They contain no pigments, although in some instances they contain a white substance, which, when chemically treated, yields a yellow hue. Like the snow and powdered glass, they owe their color to their optical properties—that is, to the reflection and refraction of the rays of light by the minute cells of which they are composed. They are derived from green, yellow, red, and blue-colored ancestors, and are the result of retrogression. In this connection the studies of white leaves by Rodrique, Laurent, and Timpe, which clearly show evidences of degeneration, are of much interest. Such leaves are thinner than normal green leaves, and consist wholly of cellular tissue, the palisade-cells being absent. Whatever impairs the vigor or vitality of the plant, as cold, impoverished soil, injury to the roots, or continued self-fertilization, will cause the floral hues to become paler, or change to white. I once transplanted a scarlet poppy when in bud, and the flowers became much smaller and changed to pure white. White flowers are most common in the cold days of early spring, and gradually become rarer toward autumn. In the arctic climate of Spitzbergen the flowers are chiefly white, and there are few yellow and red, while blue appears to fail entirely. In east Greenland the flowers are likewise chiefly white, and among 26 species there is only 1 blue. It has been observed that garden-balsams become white when ammonia is withheld from the soil, but regain their color when it is supplied. Asa Gray is reported to have said that any colored flower might revert to white, and this is undoubtedly true.

On the other hand, whatever stimulates the growth of a plant, as bright sunlight, strong manures, or crossing, increases the brilliancy of the flowers. When lowland white flowers have been cultivated in the intense light of Alpine summits they have in some cases become red. An application of nitrate of soda will increase the brilliancy of a flower; and tulips, when treated with a strong manure, flush and lose their variegated colors. The brightness of floral hues is also increased by crossing. The presence of pigments in the flowers is often correlated with its presence or absence in the leaves and stems; and it is often possible from an examination of the vegetative organs to determine beforehand whether the flowers will be white or not. The white-flowered variety of *Portulaca* has green stems, while the yellow and red varieties have red stems. A variety of *Cyclamen* with crimson flowers has the leaves purplish beneath, while the leaves of the white-flowered variety are paler and green on the under-side. The red maple has the flowers, twigs, and young leaves all crimson, while the entire plant of the wood-sorrel, including the flowers and fruit, is frequently red. In *Sedum purpureum* the petals are purple, and sometimes the entire plant; and in the stonecrop (*Sedum acre*) the foliage is yellowish green and the flowers bright yellow. It is evident that in many species the color of the flowers is determined by the pigment content of the plant as a whole.

It is easy to understand why white flowers are the most common in nature, and why they are truest to name under cultivation. Naturally florists find that they can develop

any desired color variety from a white flower more easily than
from one already containing pigments.

Nature is an excellent economist. Trees and shrubs whose
fruits are edible by man or birds usually produce their blossoms
in boundless profusion, and they are almost invariably white,
or nearly so—the two most noteworthy exceptions being the
peach and huckleberry, which have red or reddish flowers.
Among trees are the apple, pear, plum, cherries in variety, the
quince and the orange; while among shrubs are the blackberries,
blueberries, raspberries, hollies, cornels, and thorn-bushes.
There is nothing more beautiful in the floral vegetation of this
world than an apple-orchard laden with expanding blossoms.
The great masses of flowers form billowing banks of whiteness,
tinged with rose and flecked with the vivid green of the un-
folding leaf-buds, from which exhales the well-known sweet
fragrance of the apple-blossom.

> "Spring walks abroad in all the fields to-day;
>   Her touch has left the apple orchards white;
> The baby buds that waited for the May
>   Have shaken out their petals overnight;
> Against the rugged boughs they softly press,
>   Weaving in the mantle of their loveliness.
>
> Spring walks abroad with songs of life and cheer;
>   A thousand gifts she joyfully bestows;
> But all her fairest handiwork is here
>   Where orchards toss their drifts of scented snows."

Alfred Russel Wallace, who spent many years of his life in
exploring the vast forests of the Amazon and the islands of the
Malay Archipelago, declares: "I have never seen anything
more glorious than an old crab-tree in full blossom; and the
horse-chestnut, lilac, and laburnum will vie with the choicest
tropical trees and shrubs." (Fig. 109.)

Fig. 109. Button-Bush. *Cephalanthus occidentalis*
A handsome swamp-shrub with small white flowers in dense spherical heads

Fig. 110. Carrot. *Daucus Carota*

Conspicuousness is gained by the aggregation of many small white flowers in a level-topped flower-cluster.   In the lower figure the "bird-nest" formed by the cluster after it has gone to seed is shown

The largest tree-flowers known belong to the Magnolia. One Southern species has a white flower, with a purple centre, which measures ten inches across. "Their effect in early spring is grand beyond description, illuminating the whole landscape and filling the air with rich perfume." Of the five northern species four are white, and one is greenish yellow. Magnificent white flowers are likewise displayed by several species of pond-lilies; but very frequently white flowers are of small size, and conspicuousness is gained by their aggregation in masses.

Small, densely clustered white flowers standing in the same horizontal plane and affording a convenient landing-place for insects are very common in the mustard, saxifrage, carrot, honeysuckle, and aster families. This type of flower-cluster is excellently illustrated by the carrot family, or *Umbelliferæ*. (Fig. 110.) To this family belong the caraway and carrot, the wild parsnip, the water-hemlock, and the water-parsley, plants growing luxuriantly by the roadside, along the river, and in the meadow. The flowers differ very little in structure, and the species can be separated only by the aid of the mature fruit. Insects of every kind are welcome, and no other family of flowers has so large a number and variety of visitors. The nectar is fully exposed, and self-fertilization is prevented by the anthers and stigmas maturing at different times. In the *Compositæ* 126 species have either the ray or disk florets white. (Fig. 111.) In bicolored heads, or capitula, where the rays are white and the disk flowers yellow, there can be no doubt that the white rays are derived from yellow-colored ancestors. In crownbeard (*Verbesina*) all of the 5 species have yellow disks, but 1 has white and 4 yellow rays. In the genera everlasting (*Antennaria*) and cudweed (*Gnaphalium*) the yellowish-white flowers have retrograded and lost their original yellow hue.

In the orchis, pea, mint, and figwort families there occur

Fig. 111. Mayweed. *Anthemis Cotula*

Yellow centre with white rays, showing the advantage of white in rendering the flowers
conspicuous

many irregular white flowers, which have been derived from yellow, red, and blue forms, partly as the result of retrogression, and partly because of the advantage arising from a contrast of colors between closely allied species blooming at the same time. There is a white variety of the scarlet runner, and the bright-blue larkspur is sometimes white. Both white and pink flowers have been seen on a single plant of the snapdragon, and a pure-white form of the bright-red *Polygala* is sometimes found, and also of the violet-blue pickerel-weed. Everywhere there is ample evidence that a flower containing pigments may easily change to white.

Finally there are many small white flowers which are solitary, or at least not densely clustered. Some 56 such species belong to the pink family; they are low-tufted, weak herbs of a spreading or ascending habit represented by the chickweeds and sandworts. They are visited chiefly by flies and the smaller bees.

Individual white flowers may develop bright coloration during their period of blooming. The white sepals of the Christmas rose (*Helleborus niger*) regularly change to green; the flowers of a species of lantana are at first white, but later become yellow; while the corolla of the sweet-william, common pink, and *Hibiscus mutabilis* turn from white to red. In the bellflower (*Campanula*) the flowers remain white until they expand, when they change to blue. Thus under suitable conditions there may come from white flowers a great variety of colored ones. But, as has been already pointed out, white flowers are most common in primitive families in which yellow flowers are also very abundant.

COLORS OF NORTH AMERICAN FLOWERS

RED FLOWERS

There are only 257 flowers in the northeastern flora which are described as red; but as there are also 109 red-purple flowers, which should be classed with them, red flowers are thus not so rare as would at first appear. While green and yellow flowers for the most part contain solid-colored granules or plastids, red and blue flowers owe their coloration to a group of pigments dissolved in the cell-sap and called collectively anthocyanin. Considerable evidence has been obtained that the anthocyanins are derived from the yellow pigments of plants. When the anthocyanin salts are acid the flowers are red, when neutral violet, and when alkaline blue; but when the acid salt is neutralized the flowers in some instances become colorless. The color may be again restored by an acid.

Anthocyanin is very widely distributed among plants, especially among the higher or flowering plants. In early spring the new leaves of many species are suffused with it, as the red maple, the blueberry, and rhubarb; and in autumn it imparts vivid scarlet and crimson hues to the maples, huckleberries, sumacs, and blackberries. It is often abundant on the underside of floating aquatic leaves and radical leaves growing in rosettes, along stems, and in the root of the beet. It is the prevalent color of the *Coleus* and purple beech, and it adorns many fruits and berries when ripe. The anthocyanin of foliage is usually red, since the cell-sap of the vegetative organs is, as a rule, acid.

As the result of many experiments Overton found that the formation of red coloration in plants was influenced by three factors, a cell-sap rich in sugar, intense light, and low temperature. When cut stems of *Lilium Martagon* and other land-plants were placed in a 2-per-cent invert sugar-solution, red

245

anthocyanin soon appeared in the upper side of the leaves, while leaves of plants in control experiments placed in pure water remained green. Water-plants, like Hydrocharis, placed in a sugar-solution also developed red coloring in the leaves in a few days. When Overton removed his plant-cultures to the shade, the red coloring quickly disappeared, but again returned when they were exposed to bright light. Leaves, flowers, and fruits frequently display red coloring on the side exposed to direct sunlight, while the side in the shade remains green. During the summer the leaves of plants in the Alps are much oftener red-colored than in the lowlands, because the night temperature is lower and the light-intensity higher. Kerner found that the anthocyanin in plants grown in an Alpine garden at an elevation of 2,195 metres above the level of the sea was brighter-colored and more abundant than in the botanical garden at Vienna. The glumes of grasses, the leaves of stonecrops, and the pure-white petals of some flowers became red or purplish red. Winter leaves become red-colored because a lower temperature causes the sugar content to be increased at the cost of the starch.

While some ecologists regard anthocyanin as merely a by-product of the chemical activities of the cell, others, as Stahl, think that its rôle is the absorption of heat. When leaves containing anthocyanin were placed in a vessel of water the temperature of the water was raised 4 degrees higher than that of an equal quantity of water containing green leaves of the same . superficial area, both vessels being placed in the sunlight. It is readily conceivable that in early spring, when the temperature of the air is near the freezing-point, this additional heat might be a great advantage. Red coloring is seldom common in foliage containing much yellow pigment, consequently anthocyanin is rare in the birches, which have yellow leaves in spring and fall, but abundant in the same seasons in the red maple, which contains much less carrotin.

The anthocyanins are glucosides, and there are many different kinds, which are named after the flowers from which they are extracted, as that from the geranium (*Pelargonium*) is called pelargonidin, that from the larkspur (*Delphinium*) delphinidin, and that from *Malva* malvidin. The colors of flowers depend upon whether there is a small or large quantity of these pigments present, or a mixture of more than one, and also on the presence of yellow pigments.

Red has more motor power in stimulating the eye than other colors, and is consequently used commonly for railroad and other forms of signalling. It is the favorite color of savage races, and it renders persons living constantly in rooms painted bright-red nervous and excitable. Since it will excite the bull to fury and enrage the turkey-gobbler, it is not surprising to find that flowers pollinated by humming-birds and sunbirds are usually crimson or scarlet. (Fig. 112.) There are also many red pinks, lilies, and orchids, which are pollinated by red-colored butterflies, a fact which led Mueller to believe that they were likewise influenced by red coloration.

The distribution of red flowers shows that in the sequence of floral colors they are more primitive than blue, for they occur in more primitive genera and families in some of which blue flowers are unknown. The buckwheat family contains 11, the pink family 22, the rose family 19, the mallow family 13, the evening-primrose family 10, the heath family 10, and the huckleberry family 11 red-flowered species; but in northeastern America there are no blue flowers in any of these families. The flowers in the above families are regular in form and but little modified, so that it is probable that red flowers were abundant long before blue. The prevalence of red coloring in primitive groups of flowers to the exclusion of blue is due to the strongly acid condition of the cell-sap. But in the lily, pea, gentian, phlox, and mint families, which have highly specialized and often irregular flowers, both red and blue flowers are common. This second group of families are admittedly among the most recently developed in our flora, and therefore we conclude that blue flowers are of later origin than red and in many instances are derived from them.

Trees with red flowers are rare, but shrubs are common. Of trees, the peach, red buckeye, and red maple (Fig. 113) are the best-known examples. Many red-flowered shrubs occur in the rose, heath, and huckleberry families. The handsomest American shrubs are undoubtedly the Rhododendrons, Kalmias, and Azaleas, which exhibit a great variety of rose, pink, flame-colored and red shades. It is rather remarkable that among 571 species of *Compositœ* there are only 9 red flowers. On the other hand, 13 species, or one-half the mallow family, have pink or red blossoms; and there are 22 species in the pink family. In the buckwheat family the petals are wanting, but the sepals are often red and sometimes the seed-vessels, stems, and leaves. In the poppy family the flowers are crimson, scarlet, or red, and the sap is also yellow and red. The most brilliant red flower in our flora is the cardinal-flower (*Lobelia cardinalis*, Fig. 112), which is pollinated by humming-birds. There are also three flowers which are scarlet outside, but yellow within, and rarely all over; they are the wild columbine, trumpet-honeysuckle, and Maryland pinkroot, to all three of which humming-birds are common visitors.

BLUE FLOWERS

There are 325 blue flowers and 194 blue-purple flowers in the flora of northeastern America. Blue is the highest color

FIG. 112. Cardinal-Flower. *Lobelia cardinalis*

The most brilliant red flower in our flora. A humming-bird flower

in the floral world, and undoubtedly blue flowers, as a whole, were the latest evolved. They adorn the culminations in flower-building. Simple, small, regular flowers, as has already been shown, are usually white or yellow, as the water-plantains, buttercups, and fivefingers, while many red flowers are also primitive in structure. But corollas which are two-lipped, or bilaterally symmetrical, and highly modified are most frequently blue or blue-purple and are often variegated with other hues. For instance, in the buttercup family, while the buttercups are yellow, the bilateral larkspurs and monk's-hoods have blue sepals and petals. Again, in the rose family the regular rotate fivefingers are yellow and the roses are white or red, and blue flowers are entirely absent; but in the "sister family" of the pea family (*Papilionaceæ*), where the corolla is butterfly-shaped, blue and blue-purple forms prevail, which are pollinated by bees. (Figs. 20, 36, and 37.)

In blue flowers the cell-sap is neutral or alkaline, and the anthocyanin salts are violet-colored or various shades of blue and purple. There may be only a single pigment or a mixture of pigments. The color change from red to blue may be artificially illustrated by dipping a red rose in an alkaline solution, when it becomes blue, but the red hue is again restored by an acid. Many individual flowers illustrate this color transition. The flowers of the common borage (*Borago officinalis*) are at first red, but later turn blue, as do those of the lungwort (*Pulmonaria*); the corolla of the stickseed is red before expanding, but afterward becomes bright blue, and the pale-pink blossoms of the forget-me-not also soon change to blue. The gradual transition from an acid cell-sap to an alkaline one is shown by fruits, which are at first sour and red, but with maturity become sweet and blue or purple. The reverse change of color may also take place, and in a variety of perennial phlox

FIG. 113. Red Maple. *Acer rubrum*

Pistillate flowers. The entire flower is crimson, also the bud scales, twigs, and the foliage
both in spring and autumn

the flowers were a deep blue in the morning, changing to beautiful deep rose by evening.

In many species of plants the cell-sap is so nearly neutral that both red and blue flowers may be produced, or both hues may appear in the same flower. Darwin has described a hyacinth which bore on the same truss a perfectly pink and a perfectly blue flower, another truss which was blue on one side and red on the other, and also flowers which were striped longitudinally with red and blue. According to Hildebrand, red and blue cells may occur side by side in the same petal, and in the sweet violet (*Viola odorata*) there is a layer of blue cells in the epidermis, under which there is a layer of red cells in the mesophyll.

Blue anthocyanin is seldom found in yellow flowers, and plants in which the sap is very strongly acid, as the roses, may never produce blue flowers. De Candolle, therefore, concluded that yellow, red, and blue flowers could not occur in the same species; but this doctrine, to use the words of Lindley, "must now be laid up in the limbo of pleasant dreams." This supposed law is contradicted by the hyacinth, pansy, and larkspur (*Delphinium cardinale*).

Among the Monocotyledons of northeastern America (the series containing the grasses, sedges, lilies, and orchids) there are only 34 blue flowers, found chiefly in the lily and iris families. It might be supposed that the wonderful orchis family, where the flowers run riot in their strange, bizarre forms, would contain many blue flowers; but such is not the fact, and out of 6,000 species in the world there is only one, *Vanda cærulea* of India, which is blue.

Turning to the Dicotyledons, there are no blue flowers among the apetalous species. This rarity continues among the polypetalous families, for blue flowers are absent in the poppy, mustard, saxifrage, currant, rose, geranium, oxalis, spurge, holly, balsam, vine, mallow, St.-John's-wort, rock-rose, cactus, evening-primrose, ginseng, cornel families, and with three exceptions in the *Umbelliferæ*. But in three genera of the buttercup family, the columbines, larkspurs, and monk's-hoods, the violet family, and the pea family (*Papilionaceæ*) they are common. All these flowers are highly specialized, irregular, and pollinated by bees.

In the more primitive families of the *Gamopetalæ*, the series with the petals united into a corolla-tube, as the heath, primrose, olive, honeysuckle, and madder families, blue flowers are again absent. They belong chiefly to the gentian, phlox, waterleaf, borage, verbena, mint, and figwort families; while a second maximum is reached in the bellflower family (*Campanulaceæ*) and the *Compositæ*. All of these families are of comparatively recent origin, and they contain nearly 400 blue and blue-purple flowers. In the mint and figwort families, flowers of these colors are very numerous, and are often dotted, striped, or maculated with white, yellow, and red. They present the culmination of color display among flowering plants. They are mostly bee-flowers, and the majority of bee-flowers everywhere are red or blue. Of 100 species of bee-flowers in the Alps, 34 are white or yellow, and 66 red or blue. In the German and Swiss flora, 152 bee-flowers are white and yellow, and 330 red, violet, or blue. Genera adapted to bees often display a variety of colors, as violet, blue, brown, red, yellow, and white, especially when they bloom in the same locality at the same time, this contrast in hue enabling the bees to remain more easily constant to one species. Common examples are the aconites, sages, and clovers.

These highly specialized flowers often possess intricate floral mechanisms and very peculiar forms, as the skullcap, monkey-flower, snapdragon, and monk's-hood, while the nectar is so carefully concealed that few insects besides the long-tongued bees can obtain it. While in general all blue flowers are bee-flowers, not all bee-flowers have bilabiate or irregular forms. The gentian family contains many perfectly regular blue flowers which are adapted to bees. The gentians are very abundant in the Alps, and display great masses of vivid blue coloring. Huxley, while seeking health in the bracing air of these mountains, found great pleasure in studying these flowers, to an account of which his last paper was devoted. The intensity of their blue coloration has been well described by Bryant in his lines to a fringed gentian (Fig. 43):

> "Blue, blue, as if the sky let fall
> A flower from its cœrulean wall."

In the *Campanulaceæ* there are 22 blue flowers which are bell-shaped; while in the Compositæ there are many species in which the heads have blue or purple rays, as in the autumnal flowering asters. A preference for blue coloring shown by bees does not necessarily imply that blue affords them an æsthetic pleasure; but only that they recognize the signal of flowers adapted to their visits.

PURPLE FLOWERS

Red-purple flowers should be classed with red flowers, and blue-purple flowers with blue flowers (Fig. 114); but in addition to these there are in northeastern America 134 dull or lurid purple blossoms. Many of them are brownish or greenish purple, of small size and rarely visited by insects. Greenish-purple flowers which are the result of retrogression occur in the milkweed family (*Asclepiadaceæ*) and *Polygalaceæ*. From one to a few purplish flowers occur in a great number

FIG. 114. Purple Vervain. *Verbena hastata*

In parts of Iowa in favorable seasons the landscape is fairly blue with this flower, and the bees store a white honey from it

THE COLORS OF NORTHERN MONOCOTYLEDONOUS FLOWERS

Orders	Families	Yellow	White	Red	Purple	Blue	Green or Dull Color	Total
Pandanales	Typhaceæ						2	2
	Sparganiaceæ						4	4
	Naiadaceæ						42	42
Naiadales	Scheuchzeriaceæ		1				3	4
	Alismaceæ		19					19
	Vallisneriaceæ		3					3
Graminales	Gramineæ						371	371
	Cyperaceæ						334	334
Arales	Araceæ	1	2				5	8
	Lemnaceæ						11	11
Xyridales	Mayacaceæ			1				1
	Xyridaceæ	6						6
	Eriocaulaceæ						5	5
	Bromeliaceæ	1						1
	Commelinaceæ				1	11		12
	Pontederiaceæ	1	1			2		4
Liliales	Juncaceæ						47	47
	Melanthaceæ	7	10		2		5	24
	Liliaceæ	6	13	11	1	6	1	38
	Convallariaceæ	2	11	1	4		5	23
	Smilaceæ						11	11
	Hæmodoraceæ	1						1
	Amaryllidaceæ	3	3					6
	Dioscoreaceæ	1						1
	Iridaceæ	2		1		14		17
Scitaminales	Marantaceæ					1		1
Orchidales	Burmanniaceæ						1	1
	Orchidaceæ	10	18	8	14		11	61
	Total	41	82	22	22	34	857	1058

of families, as the mustard and saxifrage families. The purple trillium and Dutchman's-pipe have brown or lurid-purple hues. The custard-apple (*Asinima triloba*) is at first greenish yellow, changing to dull purple. In most instances the brown colors of flowers are due to a mixture of chlorophyll or carrotin with anthocyanin. Among the brown flowers containing two pigments are Carolina allspice (*Calycanthus*), the gooseberry, wild ginger (*Asarum*), *Adonis vernalis*, and various orchids. The black spots on the wings, or alæ, of the bean (*Vicia Faba*) contain an olive-brown pigment dissolved in the cell-sap. The spots appear black because of the flat epidermal cells. (Fig. 101.)

THE COLORS OF NORTHERN APETALOUS FLOWERS

Orders	Families	Green or Dull Color	White	Yellow	Red	Purple	Blue	Total
Piperales.......	Saururaceæ........		1					1
Juglandales.....	Juglandaceæ.......	13						13
Myricales.... {	Myricaceæ........	4						4
	Leitneriaceæ......	1						1
Salicales.......	Salicaceæ........	9		32	2			43
Fagales...... {	Betulaceæ........	7		11				18
	Fagaceæ..........	25						25
Urticales..... {	Ulmaceæ..........	3				4		7
	Moraceæ..........	6						6
	Urticaceæ........	8						8
Santalales.... {	Loranthaceæ......	2						2
	Santalaceæ........	2	2			1		5
Aristolochiales ..	Aristolochiaceæ....					10		10
Polygonales.....	Polygonaceæ.......	33	22	5	11	3		74
	Chenopodiaceæ.....	38			1			39
	Amaranthaceæ.....	16	1					17
	Phytolaccaceæ.....		1					1
Chenopodiales {	Nyctaginaceæ......		1		4	3		8
	Aizoaceæ..........		1			1		2
	Portulacaceæ......		4	3	5			12
	Caryophyllaceæ....	8	56		22	2		88
	Total.........	175	89	51	45	24		384

THE COLORS OF NORTHERN POLYPETALOUS FLOWERS

Orders	Families	Green	White	Yellow	Red	Purple	Blue	Total
Ranales	Nymphæaceæ		4	5	1	1		11
	Ceratophyllaceæ	1						1
	Magnoliaceæ		4	2				6
	Anonaceæ					1		1
	Ranunculaceæ	6	26	38	3	13	11	97
	Berberidaceæ		3	3		1		7
	Menispermaceæ	1	2					3
	Calycanthaceæ					2		2
	Lauraceæ	2		4				6
Papaverales	Papaveraceæ		5	10	6	2		23
	Cruciferæ	2	54	46	1	10		113
	Capparidaceæ		3	2	1	1		7
	Resedaceæ		1	2				3
Sarraceniales	Sarraceniaceæ			1		1		2
	Droseraceæ		4			1		5
Rosales	Podostemaceæ	1						1
	Crassulaceæ	2	2	5	2	2		13
	Saxifragaceæ	4	30	6		3		43
	Grossulariaceæ	4	6	1		2		13
	Hamamelidaceæ	1	1	1				3
	Platanaceæ	1						1
	Rosaceæ	4	35	39	13	4		95
	Pomaceæ		27		5			32
	Drupaceæ		20		1			21
	Mimosaceæ		3	1	2			6
	Cæsalpiniaceæ	2	1	7		1		11
	Krameriaceæ					1		1
	Papilionaceæ		39	33	13	88	24	197
Geraniales	Geraniaceæ		1		3	7		11
	Oxalidaceæ		1	6		1		8
	Linaceæ		1	6			2	9
	Zygophyllaceæ			2				2
	Rutaceæ	1	2					3
	Simarubaceæ	1						1
	Polygalaceæ		3	3	2	8		16
	Euphorbiaceæ	38	9	5	1			53
	Callitrichaceæ	4						4

THE COLORS OF NORTHERN POLYPETALOUS FLOWERS
(*Continued*)

Orders	Families	Green	White	Yellow	Red	Purple	Blue	Total
Sapindales	Empetraceæ					2		2
	Buxaceæ	1						1
	Limnanthaceæ		1					1
	Anacardiaceæ	7		1				8
	Cyrillaceæ	1						1
	Ilicaceæ		10					10
	Celastraceæ	3		1		2		6
	Staphyleaceæ		1					1
	Aceraceæ	5		3	1			9
	Hippocastanaceæ		1	3	1			5
	Sapindaceæ			2				2
	Balsaminaceæ			2				2
Rhamnales	Rhamnaceæ	5	3					8
	Vitaceæ	13						13
Malvales	Tiliaceæ		3					3
	Malvaceæ		4	5	13	4		26
Parietales	Theaceæ		3					3
	Hypericaceæ			22	2			24
	Elatinaceæ	4						4
	Cistaceæ	9		5				14
	Violaceæ		7	6		4	17	34
	Passifloraceæ		1	1				2
	Loasaceæ		2	3				5
Opuntiales	Cactaceæ	1		7	1	3		12
Thymeleales	Thymeleaceæ			1		1		2
	Elæagnaceæ			3				3
Myrtales	Lythraceæ	2				10		12
	Melastomaceæ				1	3		4
	Onagraceæ	3	14	24	10	6		57
	Trapaceæ		1					1
	Haloragidaceæ	6			1	6		13
Umbellales	Araliaceæ	2	3	1				6
	Umbelliferæ		58	16		1	3	78
	Cornaceæ	3	9	1		1		14
	Total	140	410	333	84	193	57	1217

THE COLORS OF NORTHERN GAMOPETALOUS FLOWERS

Orders	Families	Green	White	Yellow	Red	Purple	Blue	Total
Ericales	Clethraceæ		2					2
	Pyrolaceæ	1	7			2	1	11
	Monotropaceæ		3			1		4
	Ericaceæ		22		1	10	5	38
	Vacciniaceæ	2	10			11		23
	Diapensiaceæ		3					3
Primulales	Primulaceæ		4		11	7		22
	Plumbaginaceæ					1	1	2
Ebenales	Sapotaceæ		2					2
	Ebenaceæ				1			1
	Symplocaceæ				1			1
	Styraceæ		4					4
Gentianales	Oleaceæ	7	2				1	10
	Loganiaceæ		2		1	1		4
	Gentianaceæ		7	1	10	4	16	38
	Menyanthaceæ		2	2				4
	Apocynaceæ		2	1	1	1	2	7
	Asclepiadaceæ	7	11	3	5	13		39
Polemoniales	Convolvulaceæ	7	1		7		3	18
	Cuscutaceæ		11		1			12
	Polemoniaceæ		7		10	3	8	28
	Hydrophyllaceæ		8				10	18
	Boraginaceæ		19	6		1	17	43
	Verbenaceæ		2			2	8	12
	Labiatæ		24	4	12	48	33	121
	Solanaceæ		9	21		2	8	40
	Scrophulariaceæ		13	33	7	32	28	113
	Lentibulariaceæ			11		3	2	16
	Orobanchaceæ		1	2		2	2	7
	Bignoniaceæ		2	1	1			4
	Martyniaceæ		1					1
	Acanthaceæ				1	1	5	7
	Phrymaceæ					1		1
Plantaginales	Plantaginaceæ	14	1					15
Rubiales	Rubiaceæ	4	22	1		7	5	39
	Caprifoliaceæ		22		11	4	1	38
	Adoxaceæ	1						1
Valerianales	Valerianaceæ		5			4	1	10
	Dipsaceæ					4		4
Campanulales	Cucurbitaceæ		4	1				5
	Campanulaceæ					1	22	23
	Cichoriaceæ		8	53	5	2	5	73
	Ambrosiaceæ	15						15
	Compositæ	21	126	209	4	63	59	482
	Total	72	375	376	106	198	234	1361

CHAPTER XV

BEES AND FRUIT–GROWING

"All the forms resemble, yet none is the same as another;
Thus the whole of the throng points at a deep hidden law."
—*Goethe*.

W HILE there are many flowers with strangely bizarre
and grotesque shapes, which serve as hostelries for
bees and other insects, the majority of blossoms are
perfectly regular in form, either rotate or wheel-shaped, cup-
like or tubular. Such are the buttercup, fivefinger, strawberry, pear, apple, plum, blackberry, caraway, carrot, blueberry, goldenrod, daisy, and aster. The nectar is exposed in
many species to every passer-by, and attracts a great horde of
miscellaneous insects. Go into an orchard of Japanese plums
in early spring, and so abundant are the blossoms that they
fairly wreathe the limbs; while the air is filled with a cloud of
wild bees and flies. On the inflorescence of several species of
the carrot family (*Umbelliferæ*) more than 200 visitors have
been collected; while the goidenrods are likewise great favorites of the insect world.

In Virginia the *Ceanothus*, or New Jersey tea, is in June,
says Banks, the most attractive enchanter of insect life. Its
fragrance calls and calls till around the heads of white blossoms
there is an encircling halo of bees, flies, and beetles, which follow the enthralling odor until they rest on that bed of white.
To stand 'neath the broiling sun and watch the mazy world
of restless insect life, and to listen to the hum of a hundred
tiny wings mingled with the sharper buzz of larger species are

memories which it is pleasant to recall on many a wintry day. Banks collected 382 species of insects, a larger number than has ever been recorded for any other flower: 42 bugs (Hemiptera); 58 beetles; 165 bees, wasps, ichneumon-flies, ants, and saw-flies; and 117 flies.

Most of this great group of regular flowers, comprising tens of thousands of species, are perfect or hermaphrodite, that is, possess both stamens and pistils and can easily be self-fertilized. Are, then, the visits of insects useless, and is cross-pollination needless? Do the great mass of the higher plants depend on self-pollination, and is the much-vaunted importance of insects as pollen-carriers mythical? Or is the rank and file of the floral world as dependent on insects for pollination as are the more highly modified forms described in the previous chapters?

In the case of many plants it can easily be observed that the life cycle of the flower is divided into two distinct periods, in one of which the anthers ripen, and in the other the stigmas (dichogamy); and consequently in the absence of some agency to carry the pollen they cannot produce seed. Among entomophilous (insect-pollinated) flowers this occurs in the carrot family (*Umbelliferæ*) and among anemophilous (wind-pollinated) flowers in the sedges. In other plants the pollen does not fall on the stigmas, and still others are self-sterile; but we know so little about most wild flowers that it is often difficult to give a definite answer. There is, however, a group of trees and shrubs, and a most abundant group it is, the pollination of which has been very carefully studied by our agricultural experiment stations—our domestic fruits. So lavish is the bloom of American wild and domesticated fruits that it plays by far the most important part in the floral landscape of June, "when the white tide of bloom scuds across the land, and the gnarled apple-trees along the old stone walls are like reefs swept by surf." Let us, then, consider what the pollination of fruit-bloom can teach us.

"The continent of North America," says Hedrick, "is a natural garden. More than 200 species of tree, bush, vine, and small fruits were commonly used by the aborigines for food, not counting nuts, those occasionally used, and numerous rarities." There were whole forests of nut-trees, as the chestnut, pecan, hickory, acorn, beechnut, filbert, butternut, and nut-pine. Wild plums and cherries were abundant. Grapes, raspberries, blackberries, dewberries, gooseberries, currants, and elderberries were everywhere laden with fruit. Great areas of swamp and barren land were covered with huckleberries, blueberries, and cranberries.

Other fruits which can only be named are: "The Anonas and their kin from Florida; the native crab-apples and thorn-apples; the wineberry, the buffalo-berry, and several wild cherries; the cloudberry prized in Labrador; the crowberry of cold and arctic America; the high-bush cranberry; native mulberries; opuntias and other cacti for the deserts; the paw-paw, the persimmon, and the well-known and much-used salal and salmon berries of the west and north."

Since the days of the colonies the number of varieties of cultivated fruits have been greatly increased by hybridizing and selection. "There are now," says Hedrick, "under cultivation 11 American species of plums with 588 varieties; 15 species of grapes with 1,194 varieties; 4 species of raspberries with 28 varieties; 6 species of blackberries with 86 varieties; 5 species of dewberries with 23 varieties; 2 species of cranberries with 60 varieties; and 2 species of gooseberries with 35 varieties, or a total of 45 species with 2,014 varieties." Coville has recently shown that blueberries can be cultivated, and undoubtedly many other wild fruits will be domesticated. Improved varieties will be obtained of June-berries, elderberries, wineberries, ground-cherries, cloudberries, native mulberries, and many others. Hybridizing can multiply new forms indefinitely and yield such anomalies as the loganberry and the blackberry-dewberry.

To the list of our native fruits must be added apples, pears, plums, oranges, and other citrous fruits brought from the Old World. In California a beginning has been made in the culture of the fig, avocado, date, olive, and almond; and on a small scale the pomegranate, guava, loquat, and feijoa are being tested. The mango, a delicious fruit of which there are more than 500 varieties, has been introduced into Florida, in the southern part of which there also flourish subtropical fruits like the pineapple, banana, soursop, and cocoanut. American fruit-growing has a wonderful future before it, and the time is speedily coming when the present production, great as it is, will seem small both in quantity and variety. It is impossible to overestimate the importance of a knowledge of the pollination of fruit-bloom, and of determining whether the different varieties are self-fertile, or in the absence of insects self-sterile and unproductive. Without this knowledge their cultivation must constantly be attended by disappointment and loss.

The flowers of our common fruit-trees, the pear, apple, plum, cherry, peach, and orange are rotate, or wheel-shaped, nectariferous, and attractive to a large company of insects. On the apple there have been collected 52 species, on the pear 50, and on the sweet cherry 37, and insect visitors are equally numerous to the bloom of most other fruit trees and shrubs. Bees are most common, especially the honey-bee. Bumblebees are more often found on the blossoms of the apple than on those of the pear. There are a variety of flies of every size and a few beetles. Let us now briefly review the investigations of the experiment stations for the purpose of determining how far the productiveness of domesticated fruits is dependent on insect-pollination.

About 1875 the Old Dominion Fruit Company planted near Scotland on the James River, Va., an orchard consisting of about 22,000 standard Bartlett pear-trees. Although they always bloomed heavily and were snow-white with blossoms, they never bore a full crop; one season, when about twelve years old, they produced three-fifths of a peck per tree, whereas they should have easily yielded four or five times that quantity. Plainly there was something wrong; what was the trouble?

Waite, who was the first in America to show that many varieties of orchard-trees are self-sterile, visited this orchard in 1892, and was able by experiment to answer this question. He noticed that in some places where the Bartlett trees had died out, they had been replaced by trees of another variety, as Clapp's Favorite or Buffum. Around these trees the Bartletts were heavily laden with fruit. Mixed orchards in the vicinity also bore well. He accordingly selected a number of unopened buds and removed the stamens; and, after pollinating a part of them with pollen from Bartlett trees and a part with pollen from other varieties, enclosed them in paper bags. In the orchard at large a week after the petals had fallen the young pears all dropped off. Most of the trees were absolutely barren. Of the flowers enclosed in bags not one pollinated with Bartlett pollen had set fruit, while a large proportion of the crosses with other varieties produced pears. As there were many pollinating insects present, it is evident that had there been other varieties of pears scattered through the great orchard all of the trees would have yielded well. The Bartlett pear is largely self-sterile. (Fig. 115.)

Waite then experimented with 144 pear-trees belonging to 38 varieties. More than half of them when self-pollinated proved to be wholly or nearly self-sterile; among which were

Fig. 115. Common Pear. *Pyrus communis*

Bartlett, Anjou, Clapp's Favorite, Howell, Lawrence, and Winter Nelis. Self-fertile varieties were Angoulême, Bosc, Buffum, and Flemish Beauty. Most of the fruit, however, seems to be the result of crossing, since pollen from other varieties is prepotent over own pollen in the self-fertile varieties.

Pears produced by crossing are larger and more perfect than those which come from self-fertilization.

Like the pear many varieties of apple are self-sterile. Of 87 varieties tested by Lewis and Vincent in Oregon, 59 were found to be self-sterile; 15 were self-fertile, but gave better results when pollinated by some other variety; and 13 were partially self-sterile. Among the self-sterile varieties were Bellflower, Gravenstein, King, Rhode Island Greening, Tolman Sweet, Wealthy, and Winesap; among the self-fertile were Baldwin, Oldenburg, Shiawassee, Washington, and Yellow Newton; partially self-sterile were Ben Davis, Stark, Spitzenburg, and Yellow Transparent. In the majority of cases cross-pollination is necessary to secure a profitable crop. Cross-pollinated fruit was finer and larger, with well-developed seeds. (Fig. 116.) Do not plant in solid blocks, says Waite, but plant mixed varieties; and be sure that there are sufficient bees to pollinate the blossoms properly. (Fig. 117.)

In the *A B C of Bee Culture* the writer has given the following description of the pollination of sweet cherries: Among the orchard-trees of Oregon the cherry ranks fourth in importance, being surpassed only by the apple, prune, and pear in the order named. A poor cherry-crop affects the income of many persons. The rapid increase of the area planted with cherries has been followed by complaints that in spite of the heavy bloom there was not sufficient fruit to be profitable. In some cases new orchards have never paid expenses, while old orchards became less productive. Although sorely perplexed by these conditions, the cherry-growers, unfamiliar with the mutual relations of flowers and insects, have been slow to believe that lack of proper cross-pollination was the chief cause of the failure of their trees to set fruit. But the cherry-orchards of a decade ago were of small size and of mixed varieties; while more re-

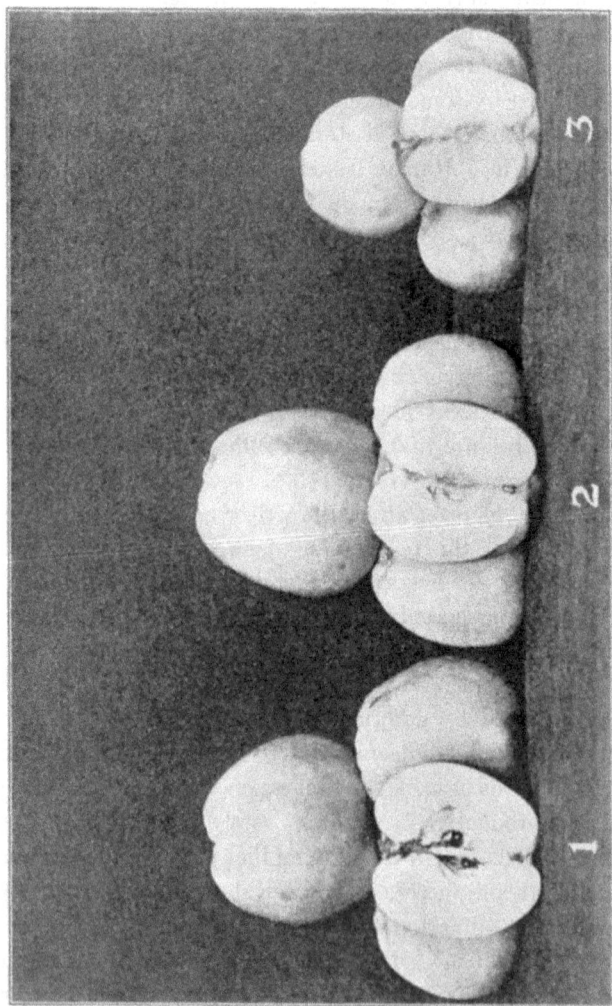

FIG. 116. Keswick Codlin Apple

1, Cross-pollinated; 2 and 3, self-pollinated. Notice the absence of seeds in the self-pollinated apples. (After Lewis and Vincent)

cently orchards of 10 to 100 acres have been planted consisting
of one or more of the standard varieties.

In order to determine the cause and remedy these failures

FIG. 117.   Apple-Blossom.   *Pyrus malus*

Gardner investigated the pollination of the sweet cherries.
Thousands of flowers were pollinated with their own pollen,
and insects excluded by bagging.   All of the 16 varieties tested
proved to be self-sterile.   Ninety per cent of the commercial
plantings consisted of the Lambert, Napoleon, and Bing, which
were not only self-sterile but intersterile, *i. e.*, each was sterile

to its own and the pollen of the other two. Napoleon when planted extensively yielded little fruit although interplanted with Lambert and Bing. But each of these varieties is effectively pollinated by Black Republican, Tartarian, and Waterhouse. Thus without cross-pollination no sweet cherries can be raised. (Fig. 88, page 181.)

The early settlers in the prairie States sometimes found native plums growing along the rivers, which were well-flavored; but when they transplanted the trees to their gardens they became unproductive. All the varieties of American plums are self-sterile, except the Robinson, and this is not wholly reliable. In the woodlands the different varieties pollinate each other. The Japanese plums are also generally self-sterile. Of the European plums a part appear to be self-sterile, and a part self-fertile; but no satisfactory experiments have been made. According to Waugh, who gave five years or more to the investigation of plum-pollination, all the species hybridize, and all the hybrids are self-sterile. The majority of plums do not bear well, and most of them set no fruit at all unless there are two or three varieties. Cross-pollination by insects is here again a necessity.

Nowhere in the world are there so many wild species of grapes as in the Eastern United States. Foreign grapes do not succeed well in this country when planted outdoors, and commercial grape-growing is, therefore, dependent on our native species. Many varieties are self-sterile. Of 169 cultivated varieties investigated by Beach in New York, 37 were wholly self-sterile, as Oneida, Eaton, Salem, and Wilder; 28 were so nearly self-sterile that the clusters were unmarketable, as Brighton, Geneva, Vergennes, and Woodruff; 104 varieties produced marketable clusters when self-fertilized, but of this number 66 had the clusters loose and only 38 yielded compact, perfect clusters, as Niagara, Agawam, Catawba, Concord, and Isabella. Nearly all the self-sterile varieties are hybrids, which cannot pollinate each other; but require pollination by self-fertile varieties in order to produce marketable clusters. A vineyard of self-sterile varieties will, therefore, produce no fruit unless there are a sufficient number of self-fertile vines planted among them to pollinate them properly. (Fig. 118.)

When the cranberry-bogs of Cape Cod and New Jersey bloom there are hundreds of level acres, which are literally covered with myriads and myriads of pinkish-white blossoms. The flowers do not furnish much nectar and, although they remain in bloom for two weeks, attract comparatively few insects. On one side of a cranberry-bog at Halifax, containing 126 acres, 3 or 4 colonies of bees were placed. This number was evidently inadequate to cover the whole field, and it was very noticeable that the crop of berries was largest nearest to the hives, and became thinner and thinner as the distance from them increased. A small piece of bog entirely screened from insects produced very little fruit. "In my travels over the United States," says E. R. Root, "I never saw a situation that demonstrated more clearly the value of bees as pollinators than did this piece of cranberry-bog."

More dissatisfaction and loss are caused among strawberry-growers from ignorance of the necessity of cross-pollination than from any other cause. A part of the plants are pistillate and a part hermaphrodite, or possess both stamens and pistils. The former remain sterile unless pollinated by insects. As the pistillate bloom is the more prolific, it is the practice in field-culture to plant three rows of pistillate to one of staminate. It is not at all rare, according to Fuller, to find perfect plants which are sterile to their own pollen, although the pollen is perfectly potent to pollinate other varieties. Unless the plants, therefore, have been tested, it is always better to plant a number of varieties in order to avoid disappointment. (Fig. 119.) The blueberries and huckleberries have pendulous, urn-shaped flowers, which are largely visited by bees. It was long supposed that blueberries could not be domesticated, but Coville has recently shown that they will grow in an acid soil. Blueberries have been produced of the size and color of Concord grapes. In Southeastern New Jersey there are thousands of acres of peaty, well-watered, pine-barrens, which are adapted to their growth. When blueberry-flowers were self-pollinated only a few berries were obtained. On some bushes not a berry matured. Neither will plants raised from cuttings taken from a single bush pollinate each other successfully, but the pollen acts as though taken from different flowers on one bush. Should a blueberry-grower set out a whole field of plants, says Coville, from cuttings from a single choice bush, his plantation would be practically fruitless. The cuttings must come from a number of not closely related bushes, and cross-pollination by bees is indispensable.

But no family can more forcibly illustrate the importance of cross-pollination than the gourd family (*Cucurbitaceæ*), which includes the cucumber, squash, pumpkin, melon, watermelon and gourd. The flowers are monœcious, that is, the stamens and pistils are in different flowers on the same plant; and in the absence of bees it is impossible for them to produce fruit unless pollinated artificially. In Massachusetts cucumbers are very extensively raised for market in greenhouses, and there are some 120 persons engaged in this industry, making use annually of more than 2,000 colonies of bees. One grower who picks 10,000 bushels requires 80 colonies, while another having some 40 acres under glass uses about the same number of hives. Without bees or hand-pollination not a cucumber would be

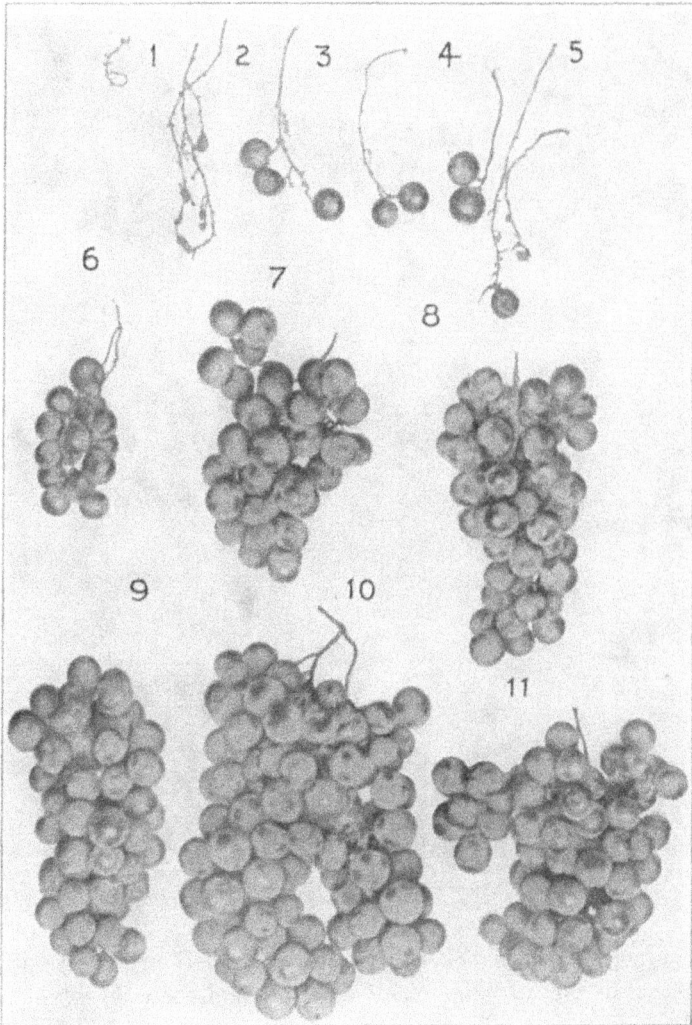

FIG. 118. Brighton Grape

Pollinated by 1, Salem; 2, Creveling; 3, Lindley; 4, pollen of another vine of the same variety; 5, self-pollinated; 6, by Nectar; 7, Jefferson; 8, Niagara; 9, Worden; 10, Vergennes; 11, Rochester. (After Beach)

Fig. 119. Strawberry. *Fragaria virginiana*

produced. Thousands of acres of cucumbers are every year grown in the fields for pickle-factories, and the crop is wholly dependent on the visits of bees. In a word, without insect-pollinators we should have no cucumbers, squashes, pumpkins, or melons. (Fig. 107, page 233.)

BEES AND FRUIT-GROWING

It is an indisputable fact that a great number of trees and shrubs will not produce fruit unless cross-pollinated by insects. At first this service was performed by our native species; but with the planting of orchards by the square mile their number became wholly inadequate to pollinate efficiently this vast expanse of bloom. This difficulty is met by the introduction of colonies of the domestic bee. No other insect is so well adapted for this purpose. In numbers, diligence, perception, and apparatus for carrying the pollen it has no equal. In orchard after orchard the establishment of apiaries has been followed by astonishing gains in the fruit-crop; and to-day it is generally admitted that honey-bees and fruit-culture must go together. "The importance of honey-bees as agents in cross-pollination," says Gardner, "cannot be overemphasized"; and one of the largest fruit-growers in New Jersey declares: "I could not do without bees. I never take a pound of their honey. All I want them to do is to pollinate the blossoms. I would as soon think of managing this orchard without a single spray-pump as without bees." The fruit-culture of the future must be largely dependent on the domestic bee, the only agency in crossing which can be controlled by man.*

Since otherwise numberless plants would produce no seed, the beneficial effects of crossing between different individuals and varieties of the same species cannot be doubted. It is by no means confined to our wild and domestic fruits, but is of very general occurrence among the higher plants. In many cases it is secured by the separation of the stamens and pistils by space, or in different flowers, as in the cone trees and many deciduous-leaved trees; or by their separation in time by one

---

* The reader who desires to follow this subject further will find it discussed at length in an article by the writer in the *A B C of Bee Culture; A Cyclopedia of Everything Pertaining to the Honey-Bee*, by A. I. and E. R. Root.

maturing before the other, as in many sedges, grasses, and in the *Umbelliferæ;* or by self-sterility, as has been illustrated in many fruit-bearing plants. That in perfect flowers the pollen ceases to be potent on its own stigma, or is even poisonous, as in certain orchids, is presumptive evidence that continuous inbreeding is injurious. It has been repeatedly shown by experiment that crossing results in the addition of new characters and increased variability, of greater fertility or the production of more and better seed, and in greater racial vigor of the offspring. In pears and apples crossed fruit was better-colored, larger, and contained many well-developed seeds, while self-fertilized fruit was much smaller and seedless or contained only vestigial seeds. Coville found that self-pollinated blueberries were smaller and later in ripening; and further examples might be multiplied indefinitely. The evil effects of inbreeding finally show themselves in decreased racial vigor, size, and fertility.

Most of the arguments against the value of crossing on examination prove to be specious. Its opponents point to the commonness of close or self-pollination; and it is assumed that the two methods must be antagonistic. In reality they are supplementary, and the great number of flowering plants, or Angiosperms, is in part due to each. Where a species is very rare and is represented by only a few individuals widely scattered, without self-fertilization it would speedily disappear, since crossing would often fail to occur. Again, where there are large areas covered with great sheets of bloom there are not sufficient insects to cross-pollinate all the flowers. It is better for a plant to be self-pollinated than not pollinated at all. In the development of both plants and animals there are not a few, which have become adapted to special locations or conditions, where they can live on almost indefinitely under self-fertilization, making no advance or even slowly retrograding.

But for plants which are actively developing and are forced into fierce competition or are compelled to meet new conditions crossing is indispensable. Darwin sowed crossed and self-fertilized seeds on the opposite sides of small pots so that there was a struggle for bare existence; and the crossed plants grew more vigorously, bloomed earlier, and more profusely, and produced more seed-capsules. Thus the inbred races tend to disappear.

Crossing is by no means confined to the individuals and varieties of the same species, but is very common between distinct species and may occur between different genera. It is rapidly coming to be regarded as an important factor in evolution. In Kerner's time more than 1,000 hybrids were known in the flora of Europe, and he fully believed that many new species originated in this way. Darwin had previously realized the possibility that hybridism might have played an important part in the history of evolution; but owing to the general belief that hybrids were almost invariably sterile he underestimated its significance, although he observed that every intermediate stage existed between complete sterility and complete fertility.

While hybrids do in general show decreased fertility, there are thousands of cases in which they multiply readily by seed. Jeffrey has recently shown that hybrids among the Angiosperms, or flowering plants, are characterized by having a part of the pollen imperfect or aborted; and, judged by this test, they are very common both among wind-pollinated and insect-pollinated plants. A great many forms which have long been regarded by systematists as perfectly good species are now recognized by their aborted pollen as hidden hybrids. They are especially abundant in the rose family among the roses, apples, pears, brambles, and hawthorns.

Among wind-pollinated flowers hybrids are very common in

the sedges (*Carex*), rushes, pondweeds, oaks, and birches. Among insect-pollinated flowers they abound among the orchids, willows, violets, and roses; while more than 1,000 species of brambles (*Rubus*) have been described in Europe, a large part of which are probably hybrids. Many of the so-called species of Cratægus, of which there seems to be no end, are the result of crossing. They are likewise abundant among the mulleins, gentians, nightshades, evening-primroses, thistles, hawkweeds, and asters.

There is not a year passes that cross-fertilization between different species does not occur on a very extensive scale; but owing to unfavorable climatic conditions or intense competition few or none even of the fertile hybrids survive. Occasionally a hybrid finds a suitable habitat and becomes a new species, and in the course of the development of the flowering plants their number has become very large. Hybrids are very variable and the great variability of the Angiosperms is doubtless due to the frequency with which crossing has taken place. Variability in turn has hastened the development of new species. Hybridism has, in the opinion of Jeffrey, clearly played a large rôle in the acceleration of the evolution of the flowering plants. It is still an active agency the investigation of which offers more promising results than any other factor in evolution. Its influence in the transformation of species may prove to be very far-reaching.

INDEX

INDEX

Bitter cress, 230
Black bees, 194, 196
Blackberries, number of varieties of, 263
Bladderwort, 19
Blister-beetle, 185, 186, 192
Bloodroot, 194
Blue, highest color in floral world, 248, 250
Blueberries, 64; domesticated, 263, 272
Blue flag, 124, 134; beetle, 124, 185
Blue flowers, 136, 138, 221–224, 248–254; coloration of, 245; number of, 248
Bluet, 203
Bombus, 122; *americanorum*, 71; *fervidus*, 52, 71, 90; *hortorum*, 71; *ternarius*, 94; *terrestris*, 71; *terricola*, 80, 100, 214; *vagans*, 80, 90, 98, 106
Bombyliidæ, 173, 174
Bonnier, 212
Borage, 60, 64, 213, 250
Botanical garden, first American, 1, 2
Bouncing bet, 151
Brambles, 278
Braun, Alexander, 20, 21
Brilliancy of flowers, 238
Brown flowers, 256
Brown, Robert, 12
Buckwheat, 116, 227; nectar-secretion of, 89, 90, 202; red-flowered, 247, 248
Bugle, 84
Bumblebee-flowers, 56, 70–88; primitive form of, 84, 86; punctures in, 80; typical wild, 78
Bumblebees, 52, 194; captured by spider, 104; maxillæ of, 96; nectaries punctured by, 96–102; pollination of red clover by, 70–72; queen, 84; species of, 71; tongue of, 74, 90
Bunchberry, 203
Burroughs, 200, 209
Butter-and-eggs, 84, 86, 232
Buttercups, 60, 92, 172, 194, 222, 224; English, 230; family, 250; colors of, 232
Butterflies, 47, 212; blue, 134; brilliant markings of, 125, 126; captured by spider, 104; flowers robbed by, 132, 134; flowers visited by, 96, 128–138; number collected on flowers, 126; number of species of, 126; punctures in plant-tissues made by, 102; red, 134, 247
Butterfly-flowers, 125 *et seq.*; Alpine, 134; best known, 128, 130; characters of, 130; colors of, 134–138; origin of, 158, 159; red, 134, 136; small number of, 132
Byturus unicolor, 188

Cabbage-butterfly, 96
Cacti, Mexican, 146
Cactus, 224
Calceolaria, 56
California, bees of, 92, 119; orange-bloom in, 115
Calla-lily, 168

Calla palustris, 164
Calliphora, 161
Calycanthus, 256
Calypso borealis, 84
Campanula, 244; *americana*, 114
Campanulaceæ, 253, 254
Canna, 228
Canterbury, 72
Caprifoliaceæ, 194
Carabidæ Lebia, 188
Caraway, 242
Cardamine, 230
Cardinal-flower, 76, 248
Carpet-beetle, 188
Carrion, beetle, 182; flower, 162; fly, 161, 212
Carrot family, 242, 261, 262
Carrotin, 228, 256; flowers owing color to, 232; light's effect on, 230, 231
Caryophyllaceæ, 130
Cassia, 200; *Chamæchrista*, 116
Castilleja, 203
Catchfly, butterfly-flower, 128; long-flowered, 151; night-flowering, 150; nodding, 151
Caterpillar, forest, 115
Catkins, staminate and pistillate, 22–28
Ceanothus, 190, 261; *americanus*, 180
Cedar-tree, 38
Centaurea cyanus, 92, 218
Century-plant, 90
Cerambycidæ, 190
Ceratopogon, 166
Cereus, 146
Cerinthe alpina, 84
Cetonia, 178, 182
Chauliognathus, 188
Checkerberry, 64
Chelone glabra, 56, 78
Chenopods, 223
Cherries, 188; ground, 194; pollination of sweet, 267–270; varieties, 269, 270
Chestnut, 232
Chickweed, 244
Children, love of flowers among, 2, 3
Chlorophyll, 226, 228, 256; elements in, 231; light's effect on, 230
Chokeberry, 180, 190
Choke-cherry, 180
Christmas rose, 244
Chrysanthemum, 236
Chrysomelidæ, 189
Chrysopsis, 234
Cinquefoils, 222
City squares, gardens in, 3
Claytonia virginica, 110
Cleistogamic flowers, 48
Clematis, garden, 196; wild, 194, 196; *Jackmanni*, 196, 219; *virginiana*, 196
Cleome, 121; *serrulata*, 112; *spinosa*, 90
Click-beetle, 189
Clintonia borealis, 227
Clover, color change in, 56; colors of, 64; nectarless, 202; red, 48, 70–72; white, 56, 119, 202; yellow, 56
Cobæa scandens, 218
Coccinellidæ, 188
Cockerell, 112

INDEX

Codling-moth, 177
Coleoptera, 178 *et seq.*
Colletes, 112, 120
Collinson, Peter, 2
Colors of flowers, 8–11, 48, 64, 65, 134–138, 200; changes in, 56, 222, 223, 230, 236, 237, 244, 250; conspicuous, 203 *et seq.;* contrast in, 206–216; determined by plant pigments, 238; distribution of, 221–224; insect preferences in, 212; variety of, 92, 208
Columbines, 64, 194, 222; bumblebee-flowers, 74, 76; nectar-secretion of, 98; punctured by bumblebees, 98, 99; reversion to regular form of, 86; wild, 76, 248
Compositæ, 110, 112, 223, 234; bee visitors of, 120, 121; beetles on, 180, 182; blue flowers of, 253, 254; butterfly visitors of, 132, 138; color in, 242; red-flowered, 248
Cone trees, 38–46
Coniferales, 40
Conifers, 40, 46
Conspicuousness, flower, advantages of, 209–220; significance of, 9–12
Coquillet, 144
Cornels, 180, 190
Cornus, 180
Corydalis, 102
Cotton, 116
Coville, 263, 272, 276
Cow-wheat, 56
Cranberries, bee-pollination of, 271; number of varieties of, 263
Cratægus, 278
Crepis aurea, 134
Crocus, moth-pollinated, 154
Cross-pollination, 12, 262; adaptations favoring, 26, 28; between different species, 277, 278; in fruit-culture, 275–278
Croton texensis, 112
Crowfoot, water, 236
Crownbeard, 242
Cucumbers, bee-pollination of, 272, 274
Cucurbita, maxima, 214; *Pepo*, 114
Cucurbitaceæ, 272
Cudweed, 242
Currant, 188; flowering, 223; golden, 237
Custard-apple, 256
Cycadeoidea, 224
Cycadophytes, 46
Cycads, 40, 46
Cyclamen, 238

Dahlia, 232; *variabilis*, 92
Daisy, 203
Dance-flies, 110, 174, 177
Dandelions, 203, 232
Daphne striata, 134
Darwin, Charles, 10–14, 17, 70, 86, 102, 158, 210, 236, 252, 277
Darwin, Francis, 12
Dates, 264
Datura arborea, 139
Davis, J. Ainsworth, 17, 150

Dead-nettle, 56, 102; nectar-guides on, 68
De Candolle, 252
Delphinidin, 247
Delphinium, 74; *cardinale*, 252; *elatum*, 74
Delpino, 13, 178
Dendrobium normale, 86
Dermestids, 188
Desmodium, 56, 198
De Vries, 150
Dewberries, number of varieties of, 263
Dianthus, arenaria, 151; *atroruber*, 134; *barbatus*, 128; *deltoides*, 128; *superbus*, 134; *sylvestris*, 134
Dicentra, 102
Dicotyledonous flowers, 221, 252
Diervilla trifida, 84
Diptera flies, 160, 171
Docks, 36
Donacia, piscatrix, 108, 185; *rufa*, 108
Dracocephalum, 84
Dragon-fly, 104
Dragon's-head, 84
Dutchman's-pipe, 168, 256

Eimer, 126
Elateridæ, 189
Elderberries, 188, 194, 200
Elecampane, 96, 132
Elm-tree, 24, 26, 28–30, 202
Emerson, 9
Empididæ, 110, 174
Empis rufescens, 19, 174
England, red clover in, 70
Entomophily, 231
Epeolus, 112
Epicauta pennsylvanica, 185
Epilobium molle, 96
Erica carnea, 132, 134
Eristalis tenax, 173
Eucalyptus, 92
Evergreen trees, 38–46
Everlasting-flower, 216, 242
Evolution, floral, 86, 222

"Facts for Darwin," 14
"Fertilization of Flowers, The," 13, 14
Fertilization. *See* Pollination
Feijoa, 264
Fig, 264
Figwort family, 48, 56, 60, 194, 242, 253; colors in, 64
Fireflies, 188
Fir-tree, 38, 40, 46
Flag, blue, 134; beetle on, 124, 185
Flies, 47, 104, 108, 160–177, 202, 264; habits of, in visiting flowers, 96; species of two-winged, 160, 161; stupid, 160; syrphid, 19, 36, 96, 160, 161, 172, 173
Florida, "Big Sawgrass" in, 116
Flower-food, surplus of, 115
Flowers, adaptations of, for effecting pollination, 68, 69; dependence of, on bee visitors, 48; two periods of life cycle of, 262; influence of, 3–6; how power of secreting nectar is lost

INDEX

INDEX

Humanity a part of nature, 8
Humboldt, 6
Humming-bird flowers, 76, 77
Humming-birds, 80, 150, 212, 247, 248
Hunt, Leigh, 7
Huxley, 71, 254
Hyacinth, 252; water, 105
Hybridizing of fruits, 263, 264; of flowering plants, 277, 278
Hybrids, 277, 278
Hydrangea, wild, 189
Hymenoptera, 160
Hyoscyamus niger, 84

Ilex glabra, 115
Impatiens biflora, 78, 80, 98, 232
Indian corn, 30, 202
Indian-pipe, 206
Indian turnip, 167, 168
Ingalls, 30
Insects, pollinating, 4, 6, 46, 160, 224, 227, 262; color preferences of, 212; difficulty of classifying, 17; value of acquaintance with, 17–19; visiting only one species of flowers, 124
Inula Helenium, 96, 132
Iowa, honey production in, 115, 116
Iris versicolor, 124, 134
Irregular flowers, bee, 48–64; white, 222, 244; yellow, 232, 234
Ivy, Boston, 228

Jack-in-the-pulpit, 167, 168
Jamestown weed, 154
Jeffrey, 277, 278
Jessamine, 232
Jewelweed, 78, 80, 232
June-bugs, 190
Juniper-tree, 46

Kalmia, 248; *latifolia*, 204
Keats, 2
Kerner, Anton, 17, 36, 114, 136, 144, 148, 182, 208, 227, 246, 277
Kerria japonica, 232
Kingsley, Charles, 3, 4
Knuth, Paul, 16, 180, 182
Kronfeld, 74

Labiatæ, 56, 64
Laburnum, 102
Lady-bugs, 188
Lamium album, 56, 68
Lantana, 223, 244
Larch-tree, 38
Larkspur, 48, 102, 132, 194, 222, 250, 252; bee, 74; bumblebee-flower, 74; regular, 86; white variety of, 244
Lathyrus odoratus, 50
Laurent, 237
Leaf-beetles, 189
Leaf-cutting bee, 52, 122
Leaf-green, 226–228
Leaf-roller moth, 140
Leaf-yellow, 228
Leaves, green, 224; red, 246; white, 237; yellow, 228
Lechea, 227

Leguminosæ, 196
Lemon, 232
Lepidoptera, 126, 159
Lewis, 267
Lily, 222; bee, 130, 131; butterfly, 130; hawk-moth, 131; moth-pollinated, 154; varieties, 130, 131, 134, 154
Linaria vulgaris, 84, 86, 232
Linden, 189
Lindley, 252
Linnæa borealis, 18, 174
Lobelia cardinalis, 248
Locust, 102
Loew, Ernst, 17, 106
Longfellow, 6
Lonicera, cœrulea, 84; *ciliata*, 84, 98; *Periclymenum*, 148; *Tartarica*, 84
Loosestrife, 194, 196; bee of, 112
Loosewort, 56
Loquat, 264
Lucilia, 161
Lungwort, 84, 174, 250
Lupine, 199–201, 206
Lupinus, 199; *subcarnosus*, 200
Lycæna, 136
Lychnis, 128, 134; evening, 151; white, 151
Lycopus europæus, 96
Lysimachia vulgaris, 112, 196

Mabie, Hamilton W., 7
Macrodactylus subspinosus, 190
Macroglossa, 146, 148
Macropis ciliata, 112, 196
Magnolia, 178, 182, 242
Maidenhair-tree, 46
Maine, bees in, 119; Perdita bee in, 112
Mallow family, red flowers of, 247, 248
Malvidin, 247
Mango, 264
Marigold, marsh, 172, 185
Maples, insect-pollination of, 21; red, 238; striped, 232
Mason bees, 194
Meadow-rue, 36
Meadow-sweet, 180
Measuring moth, 140
Medina, Ohio, apiaries at, 72
Megachile campanulæ 114; *melanophæa*, 122; *latimanus*, 52
Megarrhinus, 161
Melampyrum, 56
Melissodes, 112, 120
Meloidæ, 192
Melon, 272, 274
Memnon, statue of, 32
Mesograpta germinata, 96
Midges, 166
Milkweeds, 254; pollinating mechanism of, 177
Mint family, 48, 56, 60, 242, 253; colors in, 64
Misumena, asperata, 104; *vatia*, 104
Mitchell, Donald G., 2
Molucca balm, 84
Moluccella lævis, 84
Monarda, 134; *media*, 203; *punctata*, 121
Monkey-flower, 56, 253

INDEX

Monk's-hood, 250, 254; bumblebee flower, 74; regular, 86
Monocotyledonous flowers, colors of, 221, 257; blue among, 252
Mononychus vulpeculus, 124, 185
Mosquitoes, 161
Moth-flowers, 158, 159
Moths, two groups of, 140; nocturnal, 132; 140 *et seq.;* number listed as flower-visitors, 126; number of species of, 126; plant-tissues punctured by, 102; species of smaller, 140; tongue of, 128
Mountain-laurel, 204
Mueller, Fritz, 13, 14, 148
Mueller, Hermann, 10, 12–16, 69, 80, 98, 102, 130, 132, 160, 172, 174, 178, 210, 212, 247
Mulleins, 194, 195
Musca, 161
Mustard, 224, 236; yellow and white, 232
Myosotis, 10; *versicolor*, 223, 237

Nasturtium, garden, 78, 90, 228, 232; bumblebee-flower, 77
Nectar, depth concealed, 124, 158, 159; waste of, 115, 116; weight of, 116
Nectar-guides, 65, 68
Nectaries, variation in length of, 158, 159
Nectarless flowers, 56, 102, 202, 219
Needham, 185
Nemognatha, 182, 192
Nettles, 36, 38, 56, 223; nectar-guides on, 65, 68
New England, andrenid bees in, 120
New Jersey tea, 180; beetles on, 189, 190; insect visitors of, 261, 262
New Mexico, species of Perdita bee in, 112
New Zealand, red clover culture in, 71; flies in, 160
Nicotiana, longiflora, 154; *noctiflora*, 154
Night-blooming cereus, 146
Nightshade, 196, 234; moth-pollinated, 154; nectarless, 194
Noctuidæ, 140
Nocturnal flowers, 130, 139 *et seq.*, 146; color of, 206
Nymphæa advena, 106, 108, 230

Oak-tree, 24, 28
Odors, floral, 114; unpleasant, 161–164
Œnothera, biennis, 150; *laciniata*, 236
Old Dominion Fruit Company, 265
Olive, 264
One-sided flowers, 48–64
Orange, nectar-secretion of bloom of, 115; punctured by moth, 102
Orchids, 12, 13, 48, 159, 194, 222, 242; Madagascar, 154, 158; moth-pollinated, 154, 158; nectarless, 102 regular form in, 86
Orchis, color of, 64; blue, 136, 252; greenish or white, 132; purple-fringed, 132; showy, 84; *globosa*, 134; *Habenaria*, 132; *maculata*, 102; *morio*, 102; *spectabilis*, 84
"Origin of Species," 13
Ornithoptera, 125
Overton, 245, 246
Owlet moth, 140

Painted-cup, 76, 203
Pansy, 48, 237, 252
Panurginus, 112
Panurgus, 120
Papilio asterias, 104
Papilionaceæ, 50, 64, 250
Parnassia caroliniana, 112, 220
Parsley, water, 242
Parsnip, wild, 242
Paul, Jean, 125
Pea family, 48, 50–56, 60, 222, 242, 250; change in position and color of, 56; colors in, 64; form of, 50
Peaches, punctures in, 102
Pear, value of conspicuousness of, 212, 213; value of insect-pollination, 265; Bartlett, 265
Pear-trees, self-sterile varieties, 266
Pedicularis, 56; *sylvatica*, 84
Pelargonidin, 247
Penny-cress, 230
Perdita bee, 120; flowers visited by, 112, 121; species of, 112
Perfect flowers, 262, 276
Phillips, Doctor, 4, 116
Phlox, 132, 250
Physalis, 194
Phyteuma, 136
Phytophagous beetles, 186, 189–192
Picea mariana, 40
Pickerel-weed, bee, 105, 106; nectar-guides on, 65; insects attracted by, 106; number of bee visits to, per minute, 90; white, 244
Pigweed, 36, 223
Pinch-trap flowers, 177
Pine-trees, pollen of, 40, 45, 46; red, 40, 46; white, 38
Pinkroot, 248
Pinks, 64, 227; butterfly-flowers, 128, 130, 134; moth-pollinated, 151; red, 247; white, 234, 244
Pinus sylvestris, 182
Pinweeds, 227
Pitfall-flowers, 164, 167, 168
Plateau, Felix, 173, 212, 214, 216
Plums, American and European, 270; ground, 203; Japanese, 261, 270; punctures in, 102; number of varieties of, 263
Pogonia ophioglossoides, 84
Pollen-flowers, 194–202; anthers of, 200; devices of, for placing pollen on insects, 196–200; variety in coloration of, 200
Pollination, cross, 12, 26, 28; insect, 4, 6, 46, 160, 224, 262; self, 26, 28, 30, 224, 262; wind, 20–46
Polygala, 244
Polygaleæ, 254
Polygonum sagittatum, 96
Polypetalous flowers, colors of, 221,

284

INDEX

258, 259; rarity of blue among, 252, 253
Pomegranate, 264
Pondweed, 36
Pontederia cordata, 90, 105
Poplar-tree, 24
Poppies, 194, 224; bee visitors of, 196; colors of, 248; scarlet, 228
Portulaca, 238; *grandiflora*, 218
Potato-bug, 189
Potatoes, beetles on, 185
Primrose, 132, 134, 234; evening, 150, 232, 247
Primula acaule, 134
Prison-flowers, 164, 166
Pronuba moth, 140; habits of, 143–145
Prosopis, 121, 194
Proteaceæ, 203
Prunus virginiana, 180
Psychoda, 168
Pulmonaria officinalis, 84, 174
Pulse family, 196, 199, 201
Pumpkin, 272, 274; bee of field, 114
Punctures in corolla-tubes, 96–102
Purple flowers, 221–224, 254–256
Purslane, water, 227
Pussy-willow, change in manner of pollination of, 23, 24; insects visiting, 110
Pyrus, arbutifolia, 180; *communis*, 212

Queen of the Night, 146
Quitch-grass, 36

Ragweed, 36, 202, 223, 227
Rainfall, effect of, on red clover, 72
Raspberries, 116, 188; number of varieties of, 263
Red, effects of, 247
Red clover, 70–72, 102
Red flowers, 134, 136, 221–224, 245–248; and blue, 252; coloration of, 245–248; distribution of, 247, 248; more primitive than blue, 247
Regular flowers, 86, 261, 262
Retrogression in flowers, 227, 237, 242, 244, 254
Rhododendron, 248; *maximum*, 203
Rhodora, 9; *canadensis*, 84
Rhyncophora, 190
Ribes aureum, 223, 237
Riley, 140, 144
Rock-maple, 227, 228, 232
Rock-rose, 194
Rodrique, 237
Root, E. R., 116, 271, 275
Rose, 60, 64, 227, 228, 232, 250; beetles on, 182; insects visiting, 194; never blue, 250, 252; pollen-flower, 194, 200; no nectar secreted by, 193, 194; red, 247; yellow and white, 232
Rose-chafers, 190
Rove-beetles, 188
Rubus, 278
Rumex Acetosella, 104
Rushes, 202, 223

Sages, 56, 62, 84, 119
St.-John's-wort, 194; colors of, 234

Salix, 23, 24, 110, 121; pollen and nectar yield of, 119
Salvia, 48; *pratensis*, 84
Sand-pink, 151
Sandwort, 244
Saponaria, ocymoides, 128; *officinalis*, 151
Sarcophaga, 161
Sarcophagous beetles, 186, 188, 189
Saxifrages, dots on corolla of, 164; varieties of, 164
Scarabæids, 189, 190
Scarabæus, 190
Scarlet runner, 84, 94; punctures in, 100; white variety of, 244
Scopolia atropoides, 84
Scrophulariaceæ, 56, 64, 194
Scutellaria galericulata, 98
"Secret of Nature in the Form and Fertilization of Flowers Discovered," 9–12
Sedges, 202, 223, 262; wind-pollinated, 30–36
Sedum, acre, 238; *purpureum*, 238
Self-pollination, 224, 262; of grasses and sedges, 30; prevention of, 26, 28
Senecio, 234; *Douglasii*, 112
Shadbush, 180
Sham-nectar producers, 102
Shoe-flower, 56
Shrubs, blossoms of, 239; deciduous-leaved, 223; red-flowered, 248; yellow-flowered, 232, 234; wind-pollinated, 24
Silene, 134; *acaulis*, 128; *longiflora*, 151; *noctiflora*, 150; *nutans*, 151
Silver-winged butterfly, 96
Skullcap, 56, 253; punctures in, 98
Skunk-cabbage, 163, 164
Smilax, 227
Smith, 110
Snapdragon, 56, 62, 63, 232, 244, 254; bumblebee-flower, 78; nectar-guides on, 65; regular form of, 86
Snout-beetles, 190
Soapwort, 128
Solanaceæ, 194
Solanum Dulcamara, 196
Soldier-beetle, 188, 189
Solidago, 110, 121, 234; *bicolor*, 94, 95; *graminifolia*, 94; *juncea*, 112; *rugosa*, 94
Sorrel, 36, 104, 223, 238
Spanish bayonet, 124; moth-pollinated, 140, 143
Speedwells, 56, 172
Sphingidæ, 126, 140
Spider-plant, 90
Spiders, crab, 102–104
Spiræa, 92, 190, 194; *salicifolia*, 180
Sprengel, Christian Conrad, 9–12, 102, 210
Spring-beauty, 110
Spruce-tree, 38, 40
Squash, 232, 272, 274; value of conspicuousness of, 214
Squash-bug, 189
Stachys erecta, 56, 65
Stahl, 230, 246

285

INDEX

www.ingramcontent.com/pod-product-compliance
Lightning Source LLC
Chambersburg PA
CBHW061237220326
41599CB00028B/5455